AMERICAN ACADEMY OF ORTHOPAEDIC SURGEONS

Musculoskeletal Conditions in the United States

Allan Praemer, MA • Sylvia Furner, PhD • Dorothy P. Rice

Musculoskeletal Conditions in the United States

Steering Committee
Joseph Buckwalter, MD
Victor Goldberg, MD
Elmer Nix, MD
J. Walter Simmons, MD

Special Consultant
Jennifer L. Kelsey, PhD

Director, Center for Research
Nancy Peacock Heath, PhD

Director, Health Policy and Research
Rebecca M. Maron

Director, Communications and Publications
Mark W. Wieting

Assistant Director, Publications
Marilyn L. Fox, PhD

Art Direction and Design
Pamela Hutton

Published by the
American Academy of Orthopaedic Surgeons
222 South Prospect Avenue
Park Ridge, Illinois 60068

February 1992

First Edition

ISBN 0-89203-063-1

Authors

Allan Praemer, MA
is Research Manager with the American Academy of Orthopaedic Surgeons. He received his graduate training at Indiana University where he completed course work for his Ph.D. He has developed a number of Academy publications on orthopaedic practice from original research and from analyses of national databases. Research interests include health statistics and health care economics.

Sylvia Furner, PhD
is Assistant Professor, Epidemiology and Biostatistics in the School of Public Health, University of Illinois at Chicago. She has published a number of articles on musculoskeletal conditions, disability and cancer. Recent research has focused on morbidity and disability, especially in relation to the aging population.

Dorothy P. Rice, ScD (Hon.)
is Professor in Residence in the Department of Social and Behavioral Sciences, with joint appointments at the Institute for Health & Aging and the Institute for Health Policy Studies, University of California, San Francisco. She served from 1977 to 1982 as director of the National Center for Health Statistics, and previously served as deputy assistant commissioner for research and statistics of the Social Security Administration. Her major research interests include cost of illness studies, health statistics, impact of an aging population, and the economics of health care.

Table of Contents

Preface

I am pleased to have this opportunity to introduce readers to the American Academy of Orthopaedic Surgeons' new publication, *Musculoskeletal Conditions in the United States.* I was involved with the development of the precursor to this work, the Academy's 1984 publication, *Frequency of Occurrence, Impact, and Cost of Musculoskeletal Conditions in the United States.* The Academy's staff and expertise have now grown sufficiently to permit the revised version of the report to be developed primarily by Academy personnel.

This new publication, like its predecessor, uses several of the continuing nationwide surveys conducted by the National Center for Health Statistics for many of its statistics. The current work, however, has added a broader range of data sources, such as Health Care Financing Administration statistics, the Longitudinal Study on Aging, and newer articles from the published literature.

Additional sections on medical implants, major joint procedures, and occupational injuries have been added and several sections, especially those relating to costs of musculoskeletal conditions, neoplasms of bone and connective tissue, and health care utilization, have been expanded.

I am sure that persons engaged in research and delivery of care involving musculoskeletal conditions will find this book most useful as a reference tool and data source. The enormous impact of musculoskeletal conditions is not generally appreciated. This book should help highlight the significance of musculoskeletal diseases and injuries, and their effects on the health and well-being of the American population.

Jennifer L. Kelsey, PhD

Professor of Health Research and Policy and
Chief, Division of Epidemiology

Department of Health Research and Policy

Division of Epidemiology

Stanford University School of Medicine
Stanford, California

SECTION 1

Overview

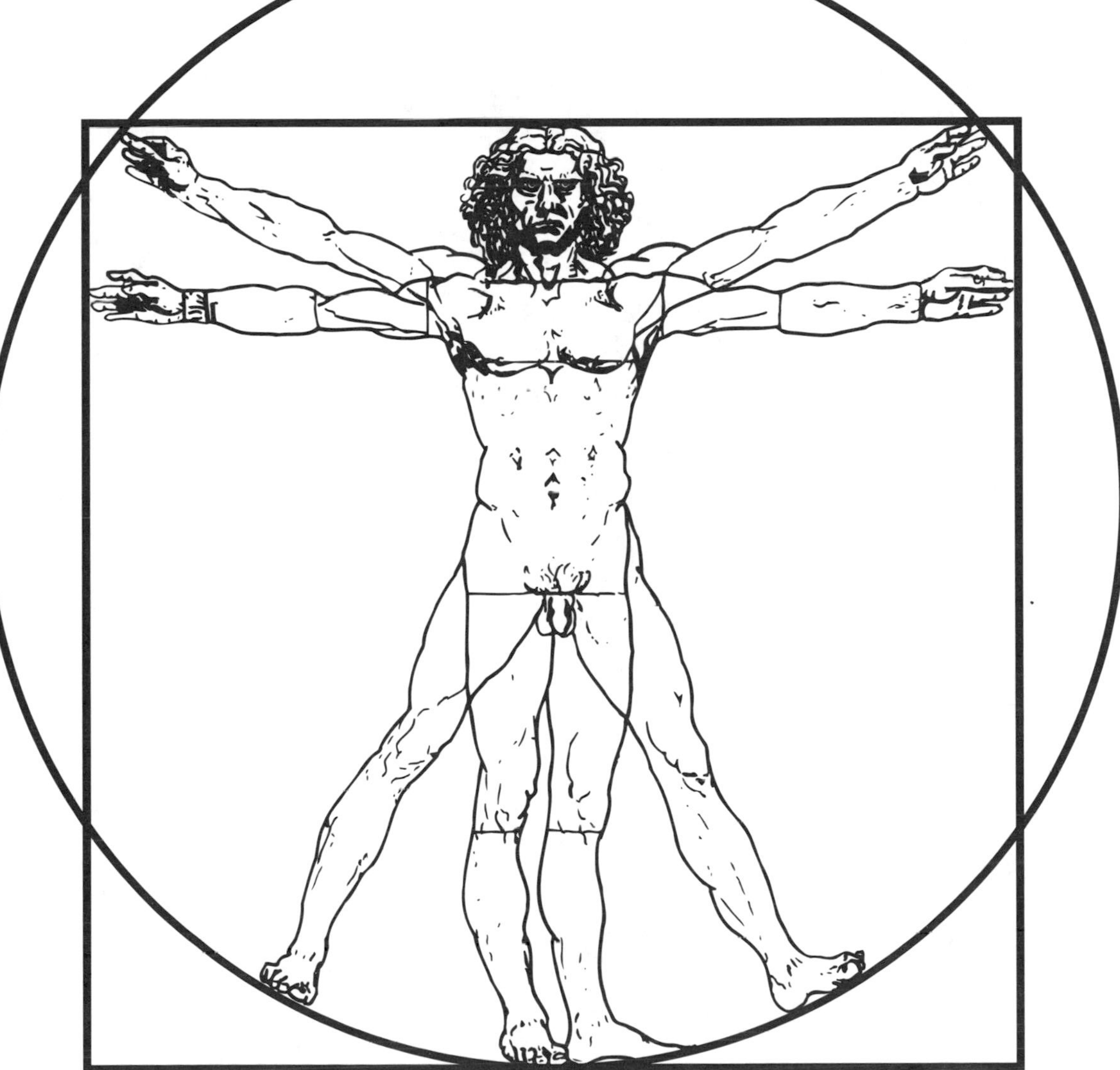

Overview

Musculoskeletal conditions are among the most frequently occurring medical conditions, and they have a substantial impact on the health and quality of life of the population as well as on the use of health care resources.

Three categories of impact

The impact of musculoskeletal conditions includes three major categories: (1) the physical and social impact resulting from increased pain, limitations on mobility and activities of daily living, loss of independence and a reduced quality of life; (2) direct expenditures for the diagnosis and treatment of musculoskeletal diseases and trauma; and (3) the indirect economic loss associated with reduced participation in the labor force, as well as lost productivity and wages that result from activity limitations induced by musculoskeletal impairment and disability.

The impact of musculoskeletal conditions is, in large part, a function of their prevalence in the population. Among impairments identified in the National Health Interview Survey (1988), musculoskeletal impairments are the most frequently reported, with approximately 29.9 million reported in the United States in 1988. Musculoskeletal impairments are widely distributed in the population and are also the most frequently reported among both sexes and among major racial groups (Tables 1 and 2). An impairment is a chronic or permanent defect representing a decrease or loss of ability to perform various functions. By age group, musculoskeletal impairments are the most prevalent through the 45 to 64 age group (Table 3).

Prevalence

In the United States population, musculoskeletal impairments occur at a rate of approximately 124.0/1,000 persons. Back or spine impairments are the most frequently reported subcategories and represent 51.7% of musculoskeletal impairments (Figure 1).

Musculoskeletal impairments are reported at virtually the same rate among women (124.3/1,000) as men (123.6/1,000) (Table 4). Although the difference in rate between men and women is not statistically significant, rates vary substantially by race and age group. Among whites, the rate of musculoskeletal impairments is approximately 26% higher than

Table 1: Prevalence of Selected Impairments, United States, 1988 By Gender

Type of Impairment	Estimated Number of Impairments (in millions) Total	Males	Females
Musculoskeletal Impairments (except paralysis or amputation)	**29.866**	**14.422**	**15.444**
Back or Spine	**15.431**	**6.701**	**8.730**
Lower Extremity or Hip	**11.126**	**5.892**	**5.234**
Upper Extremity or Shoulder	**3.309**	**1.829**	**1.480**
Paralysis, Complete or Partial	**1.296**	**0.615**	**0.681**
Absence of Major Extremities	**1.514**	**1.292**	**0.222**
Hearing Impairments	**21.864**	**12.296**	**9.468**
Visual Impairments	**8.365**	**5.154**	**3.211**
Speech Defects	**2.640**	**1.640**	**1.000**

Source: National Center for Health Statistics, National Health Interview Survey, 1988.

Table 2: Prevalence of Selected Impairments, United States, 1988 by Race

Type of Impairment	Estimated Number of Impairments (in millions)			
	Total	White	Black	Other
Musculoskeletal Impairments (except paralysis or amputation)	**29.866**	**26.251**	**3.013**	**0.463**
Back or Spine	**15.431**	**13.957**	**1.137**	**0.336**
Lower Extremity or Hip	**11.126**	**9.376**	**1.524**	**0.088**
Upper Extremity or Shoulder	**3.309**	**2.918**	**0.352**	**0.039**
Paralysis, Complete or Partial	**1.296**	**1.110**	**0.149**	**0.036**
Absence of Major Extremities	**1.514**	**1.317**	**0.141**	**0.056**
Hearing Impairments	**21.864**	**20.204**	**1.388**	**0.272**
Visual Impairments	**8.365**	**7.532**	**0.694**	**0.139**
Speech Defects	**2.640**	**2.144**	**0.433**	**0.064**

Source: National Center for Health Statistics, National Health Interview Survey, 1988.

Table 3: Prevalence of Selected Impairments, United States, 1988 By Age

Type of Impairment	Total	Less than 18	18-44	45-64	65-74	75-84	85 & over
	Estimated Number of Impairments (in millions)						
Musculoskeletal Impairments (except paralysis or amputation)	29.866	1.890	14.920	7.942	2.994	1.667	0.453
Back or Spine	15.431	0.714	8.295	4.105	1.333	0.780	0.203
Lower Extremity or Hip	11.126	1.104	5.214	2.734	1.238	0.600	0.236
Upper Extremity or Shoulder	3.309	0.072	1.411	1.103	0.423	0.287	0.014
Paralysis, Complete or Partial	1.296	0.080	0.280	0.439	0.366	0.096	0.034
Absence of Major Extremities	1.514	0.010	0.476	0.420	0.429	0.143	0.036
Hearing Impairments	21.864	1.078	5.021	6.725	4.807	2.963	1.269
Visual Impairments	8.365	0.579	3.010	2.174	1.184	0.902	0.516
Speech Defects	2.640	1.151	0.728	0.380	0.245	0.072	0.064

Source: *National Center for Health Statistics, National Health Interview Survey, 1988.*

Figure 1: Distribution of Musculoskeletal Impairments: United States, 1988

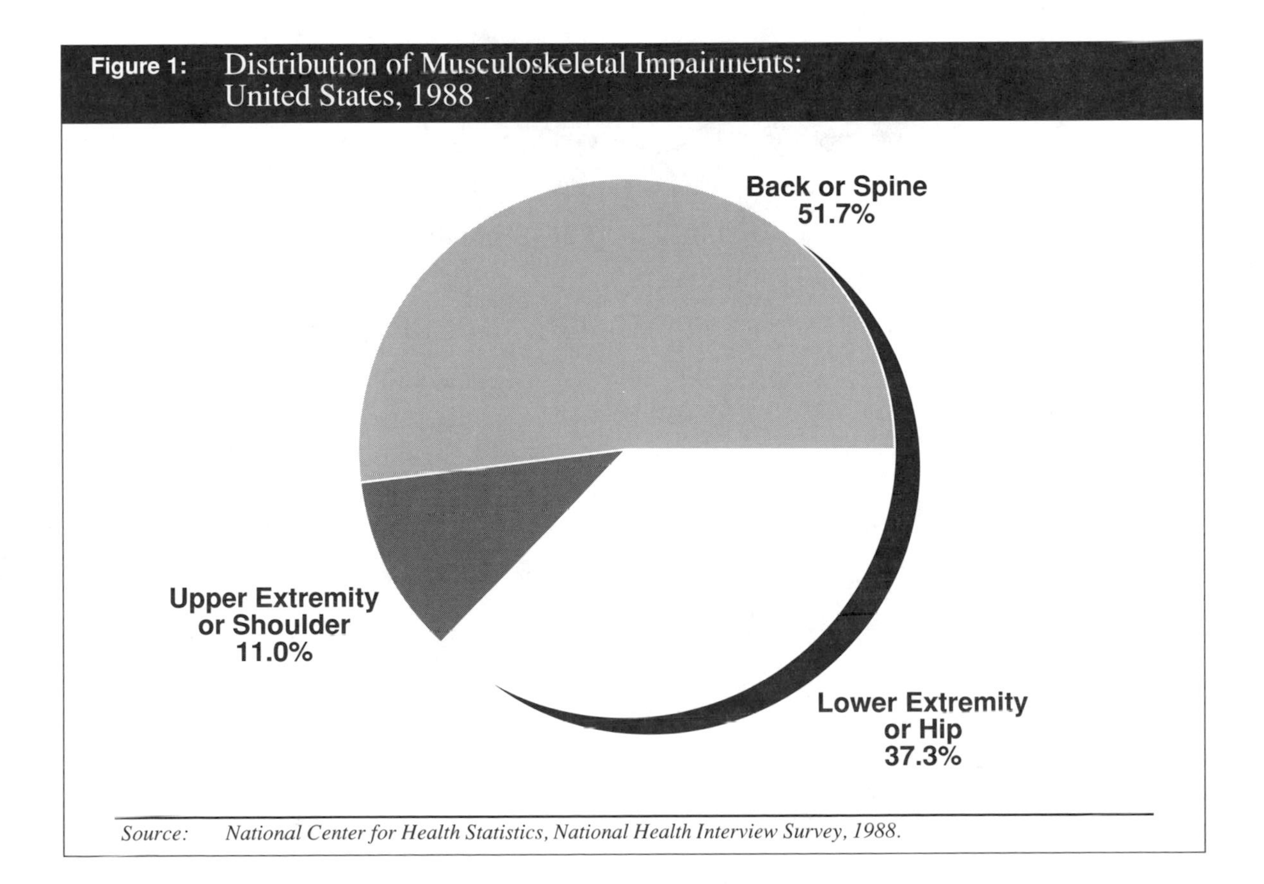

Source: *National Center for Health Statistics, National Health Interview Survey, 1988.*

Table 4: Prevalence of Musculoskeletal Impairments by Major Population Subgroups: United States, 1988

		Impairments/1,000 Population			
		All Musculo-skeletal	Back or Spine	Lower Extremity or Hip	Upper Extremity or Shoulder
Total		124.0	64.1	46.2	13.7
Gender:	Male	123.6	57.4	50.5	15.7
	Female	124.3	70.3	42.1	11.9
Race:	White	129.1	68.7	46.1	14.4
	Black	102.5	38.7	51.9	12.0
Age:	0-17 years	29.7	11.2	17.4	1.1
	18-44 years	144.8	80.5	50.6	13.7
	45-64 years	174.3	90.1	60.0	24.2
	65 & over	178.3	80.8	72.3	25.2
	65-74 years	170.4	75.9	70.5	24.1
	75-84 years	186.3	87.2	67.0	30.1
	85 & over	208.9	93.6	108.8	6.5

Source: National Center for Health Statistics, National Health Interview Survey, 1988.

among blacks.Spine or back impairments among whites (68.7/1,000) exceed the rate for blacks (38.7/1,000).

Prevalence rates of musculoskeletal impairments, as expected, increase with age. Within the adult population (comparing the 18 to 44 and 65 and older age groups), 64% of the increase results primarily from a higher prevalence of lower extremity or hip impairments (Figure 2). Among those 85 and older, the prevalence rate of musculoskeletal impairments is 208.9/1,000, with lower extremity or hip impairments accounting for 52% of musculoskeletal impairments.

Limitation of activity

Musculoskeletal impairments are a leading cause of activity limitation (Table 5) and in 1988 alone resulted in over 382.2 million restricted activity days, including 124.0 million bed days. Almost half (48.5%) of restricted activity days resulted from back or spine impairments, 38.5% from lower extremity or hip impairments. The distribution of bed days is more concentrated, with 67.1% attributable to back or spine impairments, 25.2% to lower extremity or hip impairments, and 7.7% to upper extremity impairments.

Figure 2: Rate of Musculoskeletal Impairments by Age: United States, 1988

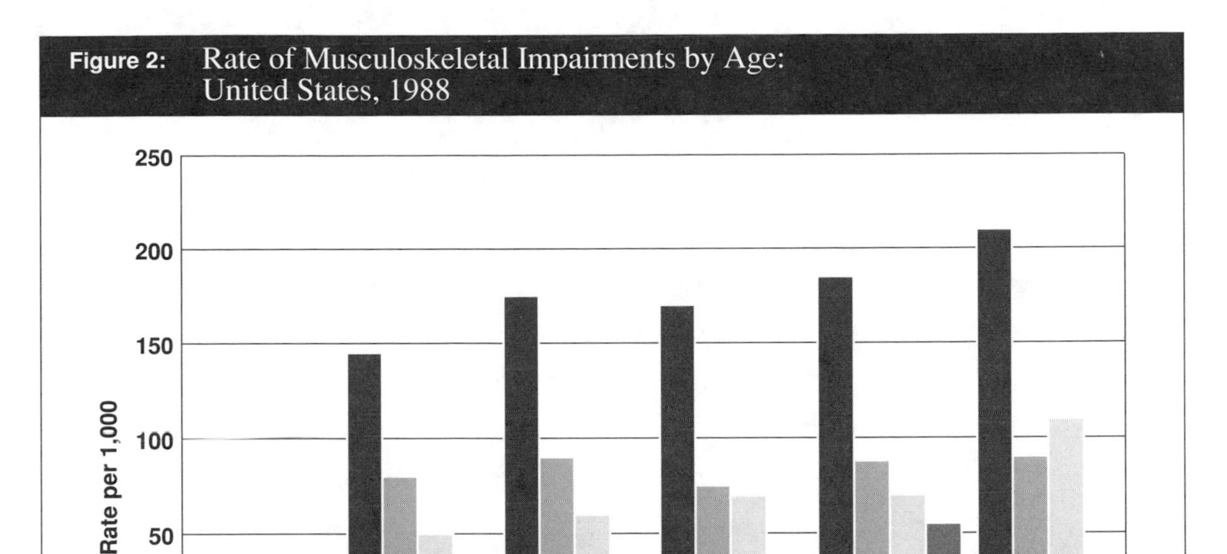

Source: *National Center for Health Statistics, National Health Interview Survey, 1988.*

Table 5: Number of Restricted Activity Days and Bed Days for Selected Impairments United States, 1988

(Number of Days in thousands)

Type of Impairment	Restricted Activity Days	Restricted Activity Days/ Impairment	Bed Days	Bed Days/ Impairment
Hearing Impairment	**10,357**	**0.5**	**989**	**0.045**
Visual Impairment	**46,213**	**5.5**	**15,614**	**1.9**
Speech Impairment	**11,462**	**4.3**	**2,118**	**0.8**
Musculoskeletal Impairments	**382,203**	**12.8**	**124,003**	**4.2**
Back or Spine	**185,542**	**12.0**	**83,165**	**5.4**
Lower Extremity	**147,228**	**13.2**	**31,284**	**2.8**
Upper Extremity	**49,433**	**14.9**	**9,554**	**2.9**
Paralysis	**51,308**	**39.6**	**8,435**	**6.5**
Absence of Major Extremities	**22,909**	**15.1**	**3,932**	**2.6**

Source: *National Center for Health Statistics, National Health Interview Survey, 1988.*

Musculoskeletal impairments result, on average, in 12.8 days of restricted activity, including 4.2 bed days (Table 6). Restricted activity days are highest for upper extremity impairments (14.9 days); bed days from back or spine impairments (5.4 days). Restricted activity days per impairment are higher among males, blacks and those 65 and older, primarily resulting from longer restricted activity levels for back and upper extremity impairments.

Bed days show a similar pattern, with higher averages for males and blacks but, in contrast to restricted activity days, those younger than age 65 show a lower average number of bed days.

Nursing home residents

The impact of musculoskeletal conditions is not restricted to the community-living population. Diseases of the musculoskeletal system and connective tissue disorders rank fourth by major system or disease category in frequency among nursing home residents and were indicated for 429,000 nursing home residents (78,000 males and 351,000 females) (Table 7). Arthritis and rheumatism accounted for 63.4% of

Table 6: Restricted Activity Days and Bed Days/Impairment for Musculoskeletal Impairments: United States, 1988 by Major Population Subgroup

	Restricted Activity Days/Impairment				*Bed Days/Impairment*			
	All Musculo-skeletal	Back or Spine	Lower Extremity or Hip	Upper Extremity or Shoulder	All Musculo-skeletal	Back or Spine	Lower Extremity or Hip	Upper Extremity or Shoulder
Total	**12.8**	**12.0**	**13.2**	**14.9**	**4.2**	**5.4**	**2.8**	**2.9**
Gender								
Male	**13.5**	**14.5**	**11.6**	**16.2**	**4.5**	**5.9**	**3.1**	**3.8**
Female	**12.1**	**10.1**	**15.1**	**13.3**	**3.8**	**5.0**	**2.5**	**1.8**
Race								
White	**11.5**	**10.1**	**13.3**	**12.7**	**3.3**	**4.2**	**2.7**	**1.1**
Black	**21.4**	**31.8**	**14.9**	**16.4**	**8.0**	**15.4**	**3.7**	**2.4**
Age								
Less than 65	**12.0**	**11.5**	**11.3**	**16.3**	**4.3**	**5.4**	**2.9**	**3.6**
65 & over	**17.0**	**15.0**	**21.5**	**10.2**	**3.4**	**5.3**	**2.4**	**0.3**

Source: National Center for Health Statistics, National Health Interview Survey, 1988.

Table 7: Prevalence of Major Impairments Among Nursing Home Residents: United States, 1985 by Gender and Age Group* *(in thousands)*

	Total	Under 65	Total over 65	65-74	75-84	85 & over
Diseases of the Circulatory System	1,521	73	1,448	187	519	742
Male	376	35	341	66	141	134
Female	1,145	38	1,107	122	378	608
Mental Disorders	690	124	566	114	228	223
Male	209	65	144	46	58	40
Female	481	60	422	68	170	183
Diseases of the Nervous System and Sense Organs	509	81	429	84	166	179
Male	165	41	124	34	50	40
Female	344	40	305	50	116	139
Diseases of the Musculoskeletal System	429	19	410	46	142	222
Male	78	7	71	15	26	30
Female	351	12	339	32	115	192
Arthritis or Rheumatism	272	7	265	26	88	151
Male	40	1	39	7	13	20
Female	231	5	226	19	76	132
Osteoporosis	49	1	48	3	16	29
Male	4	0	4	0	1	3
Female	45	1	45	3	16	26
Injuries and Poisonings	101	11	90	7	28	54
Male	21	9	12	2	6	5
Female	80	2	78	5	23	50
Hip Fracture	39	2	37	2	11	24
Male	6	1	5	1	2	2
Female	33	0	33	1	9	22
Other Fracture	36	2	34	3	10	20
Male	4	1	3	1	1	1
Female	31	1	31	2	9	19

**Number of all-listed diagnoses at time of survey.*

Source: *National Center for Health Statistics, National Nursing Home Survey, 1985.*

these musculoskeletal diseases, including 65.8% of musculoskeletal conditions among women and 51.3% among men.

Osteoporosis accounted for an additional 11.4% of musculoskeletal disease diagnoses and was diagnosed primarily among women.

The category "injury and poisonings" accounted for an additional 101,000 diagnoses. A large majority of these were fractures, with hip fractures accounting for 38.6% and other fractures an additional 35.6% of diagnoses in this category. Most hip fracture and other fracture diagnoses occurred among women (84.6% and 86.1%, respectively).

Prevalence rates for selected musculoskeletal conditions among nursing home residents are indicated in Figure 3. Among all nursing home residents, the prevalence rate for musculoskeletal diseases is approximately 250/1,000 residents. The rate is substantially higher among female (280/1,000) than among male residents (163/1,000).

Figure 3: Prevalence Rates* of Selected Musculoskeletal Diseases for Nursing Home Residents, by Gender and Age, 1985

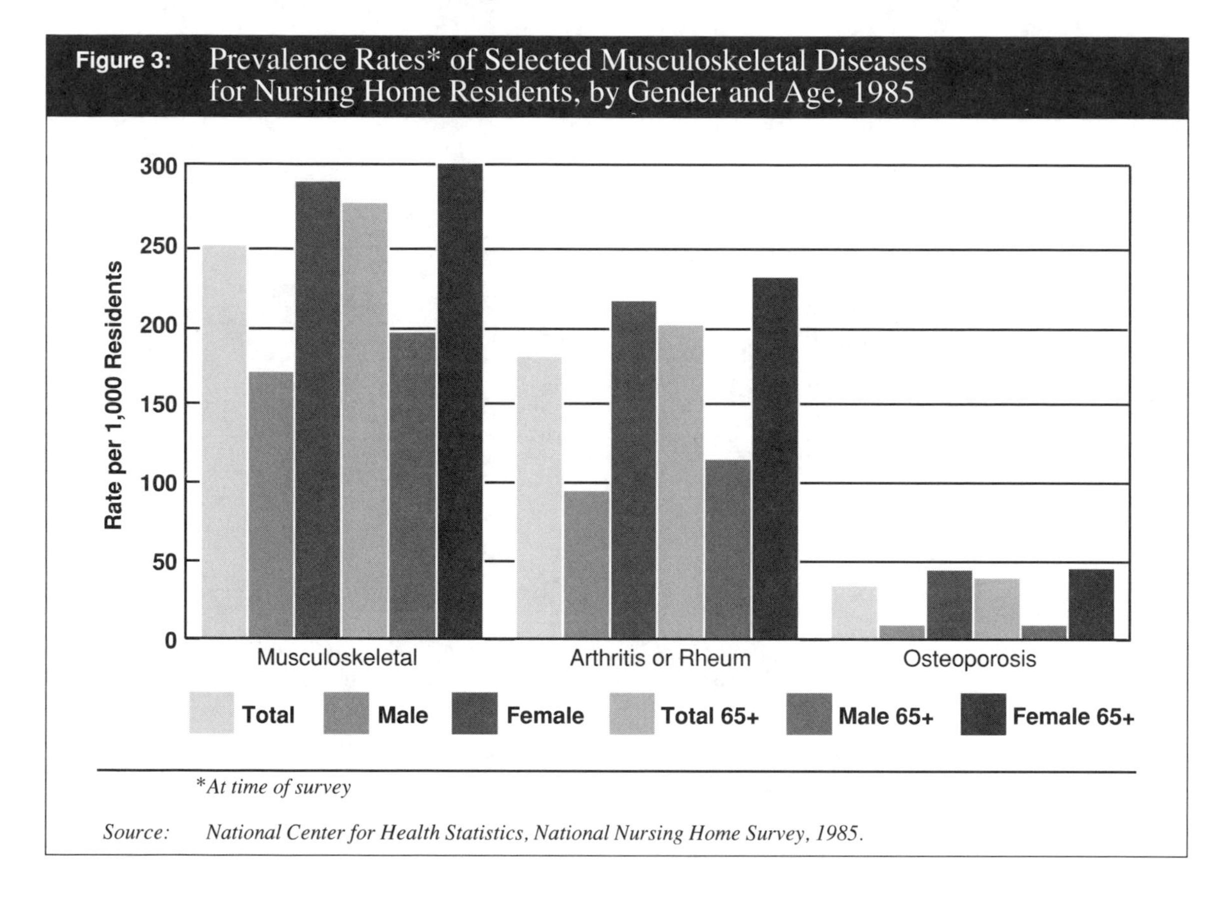

*At time of survey

Source: *National Center for Health Statistics, National Nursing Home Survey, 1985.*

Injury

Musculoskeletal injuries occur frequently and result in significant disability and use of health care resources. Over one-half of all injuries reported in the National Health Interview Survey represent injuries to the musculoskeletal system (Table 8). There were approximately 32 million musculoskeletal injuries in the United States during 1988, an incidence rate of 132.8 per thousand population (Table 9). Incidence rates were higher for males than females (158.4/1,000 vs. 108.9/1,000, respectively).

Use of health care resources

Musculoskeletal conditions result in a significant use of health care resources. In 1988, there were approximately 3.5 million hospitalizations in short-stay hospitals for musculoskeletal conditions, accounting for

Table 8: Number of Persons Injured: United States, 1988 by Gender and Age

(thousands of persons)

	All Injuries	Musculoskeletal Injuries	Fractures	Sprains and Dislocations
Total	**57,702**	**32,001**	**6,535**	**14,771**
Male	**31,758**	**18,476**	**3,643**	**7,697**
Female	**25,945**	**13,526**	**2,891**	**7,074**
Less than 18 years	**18,191**	**8,684**	**1,961**	**4,034**
18-44 years	**26,950**	**16,201**	**2,796**	**7,979**
45-64 years	**6,859**	**4,261**	**1,078**	**1,909**
65 years & over	**5,703**	**2,855**	**700**	**848**

Source: *National Center for Health Statistics, National Health Interview Survey, 1988.*

Table 9: Incidence of Persons Injured: United States, 1988 by Gender and Age

(Persons injured/1,000 population)

	All Injuries	Musculoskeletal Injuries	Fractures	Sprains and Dislocations
Total	**239.5**	**132.8**	**26.3**	**61.3**
Male	**272.2**	**158.4**	**31.2**	**66.0**
Female	**208.8**	**108.9**	**23.3**	**56.9**
Less than 18 years	**286.2**	**136.6**	**30.8**	**63.5**
18-44 years	**261.5**	**157.2**	**27.1**	**77.4**
45-64 years	**150.5**	**93.5**	**23.7**	**41.9**
65 years & over	**198.8**	**99.5**	**24.4**	**29.6**

Source: *National Center for Health Statistics, National Health Interview Survey, 1988.*

12.8% of all hospitalizations. Musculoskeletal conditions ranked second only to diseases of the circulatory system in overall frequency (Figure 4).

Figure 4: Inpatients Discharged from Short-Stay Hospitals by Category of First Listed Diagnosis: United States, 1988*

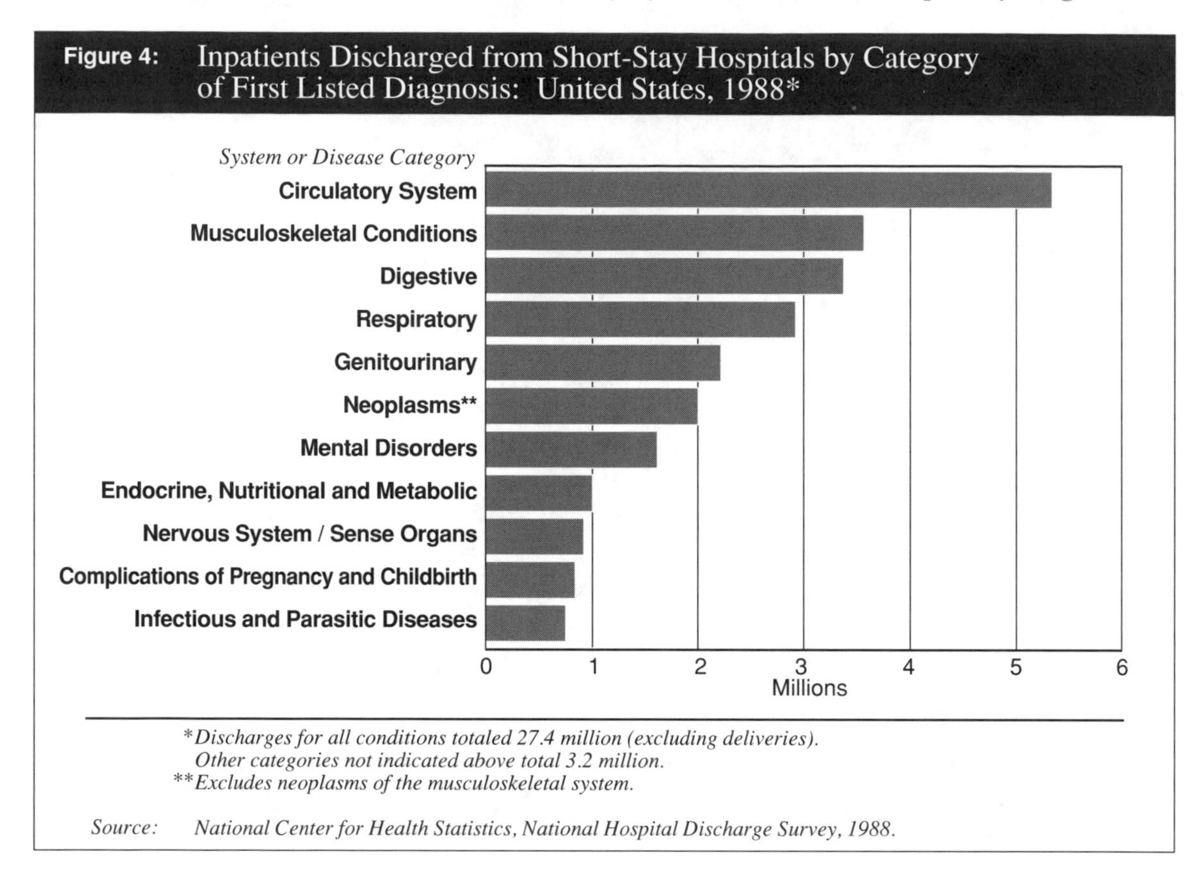

*Discharges for all conditions totaled 27.4 million (excluding deliveries). Other categories not indicated above total 3.2 million.
**Excludes neoplasms of the musculoskeletal system.

Source: National Center for Health Statistics, National Hospital Discharge Survey, 1988.

Examining musculoskeletal conditions by major category (Table 10) indicates that musculoskeletal diseases and connective tissue disorders account for the largest number of hospitalizations (1,647,000), including 459,000 for arthropathies and related disorders and 417,000 for intervertebral disk disorders. Fractures resulted in 899,000 hospitalizations, with the remaining trauma categories, dislocations and sprains, and other injuries, an additional 526,000.

Musculoskeletal diseases and connective tissue disorders account for almost half (46.9%) of hospitalizations associated with musculoskeletal conditions. Trauma (fractures, dislocations and sprains, and other injuries) accounts for an additional 40.5% (Figure 5). Complications (infection and inflammatory reactions to internal orthopaedic or prosthetic devices, implants or grafts) account for about 5%, with smaller percentages resulting from neoplasms (3.9%) and congenital anomalies (1.2%).

Table 10: Hospitalizations Resulting from Musculoskeletal Conditions: United States, 1988 by Aggregate Category*

Category	Percent	Number	Subcategory Number
Musculoskeletal Diseases and Connective Tissue Disorders	46.9%	1,647,000	
Arthropathies and Related Disorders			459,000
Intervertebral Disk Disorders			417,000
Other Back Disorders			178,000
Fractures	25.6%	899,000	
Fracture of Neck of Femur			254,000
Dislocations and Sprains	7.3%	258,000	
Sprains and Strains of the Back			97,000
Other Injuries	7.6%	268,000	
Complications or Reactions**	4.9%	172,000	
Congenital Anomalies	1.2%	41,000	
Neoplasms	3.9%	136,000	
Other Musculoskeletal Conditions	2.6%	92,000	
Total, All Musculoskeletal Conditions		**3,513,000**	

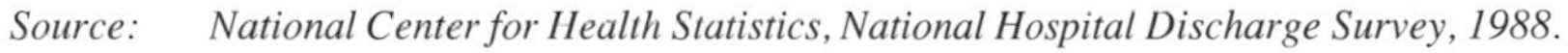

**First listed diagnosis for inpatients discharged from short stay hospitals.*
***Mechanical, other complication or infection and inflammatory reaction to internal orthopaedic or prosthetic device, implant or graft.*

Source: *National Center for Health Statistics, National Hospital Discharge Survey, 1988.*

Figure 5: Distribution of Hospitalizations Resulting from Musculoskeletal Conditions: United States, 1988 by Aggregate Category

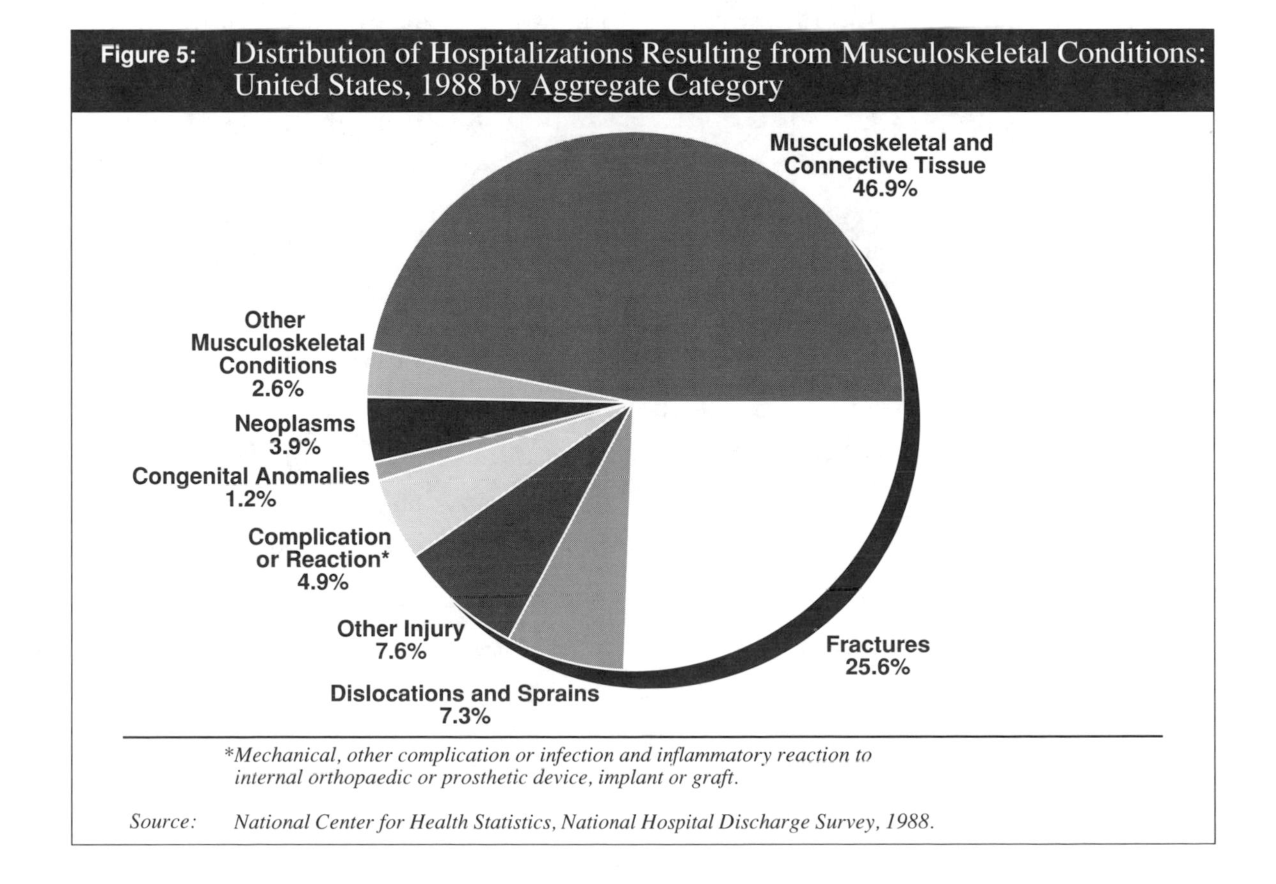

**Mechanical, other complication or infection and inflammatory reaction to internal orthopaedic or prosthetic device, implant or graft.*

Source: *National Center for Health Statistics, National Hospital Discharge Survey, 1988.*

Age group differences

The distribution of musculoskeletal conditions resulting in hospitalization varies by age group (Table 11 and Appendix A). In the younger-than-18 age group, trauma and congenital anomalies make up a comparatively high percentage of musculoskeletal conditions. Among those 18 to 44, musculoskeletal diseases and connective tissue disorders increase in importance and primarily reflect increases in (1) internal derangement of the knee and other joints and (2) intervertebral disk and other back disorders (Appendix A). Musculoskeletal diseases and connective tissue disorders reach their highest relative importance in the 45 to 64 age group. This results from increased hospitalizations for (1) arthropathies and related disorders, (2) spondylosis and allied disorders, and (3) continued high levels of disk and other back disorders.

Among those 65 and older, fractures account for an increasing proportion of musculoskeletal hospitalizations. Fractures account for 35.1% of musculoskeletal related hospitalizations among those 65 and older, including 39.1% in the 75 to 84 age group and 60.5% in the 85 and older age group.

The use of hospital resources helps illustrate the disproportionate impact of musculoskeletal conditions on the aged population, especially women 65 and older. The 65 and older age group, for example, accounts for 31.4% of hospitalizations for musculoskeletal conditions but makes up only 12.4% of the United States population. Women over 65 account for 21.4% of hospitalizations for musculoskeletal conditions, almost three times greater than their representation in the United States population (7.4%).

The overall hospitalization rate for musculoskeletal conditions among the United States civilian persons is 144.2 hospitalizations/10,000 persons (Table 12) and is higher among men (147.8) than women (140.8). The rate overall varies directly with age and, among those 65 and older, is more than two-and-one-half times the average among all age groups. The increase is more dramatic within the 65 and older group. The rate of hospitalizations for musculoskeletal conditions increases from 283.0/10,000 persons for those age 65 to 74 to 621.4/10,000 for those age 85 and older.

Although, as indicated in Table 12, the overall hospitalization rate for musculoskeletal conditions is higher among men, women use hospital resources more in the 65 and older categories. Among those age 65 and older, women are one-and-one-half times more likely than men to be

Table 11: Hospitalizations Resulting from Musculoskeletal Conditions: United States, 1988*

	Percent Distribution by Age Group							
	Total	Less than 18	18-44	45-64	65 & over	65-74	75-84	85 & over
Musculoskeletal Diseases and Connective Tissue Disorders	**46.9**	**25.9**	**48.1**	**57.5**	**43.1**	**51.9**	**39.1**	**28.7**
Fractures	**25.6**	**37.2**	**20.2**	**16.6**	**35.1**	**22.9**	**39.1**	**60.5**
Dislocations and Sprains	**7.3**	**5.5**	**11.4**	**7.6**	**3.0**	**3.9**	**2.6**	**1.5**
Other Injuries	**7.6**	**13.9**	**13.0**	**3.5**	**3.0**	**2.8**	**3.3**	**2.7**
Complications or Reactions**	**4.9**	**2.9**	**3.0**	**5.7**	**6.9**	**7.3**	**8.4**	**2.2**
Congenital Anomalies	**1.2**	**8.4**	**0.5**	**0.6**	**0.2**	**-**	**-**	**-**
Neoplasms	**3.9**	**3.6**	**1.8**	**5.0**	**5.9**	**7.3**	**4.7**	**3.5**
Other Musculoskeletal Conditions	**2.6**	**2.5**	**1.8**	**3.6**	**2.8**	**3.7**	**2.6**	**0.7**

**First listed diagnosis for inpatients discharged from short stay hospitals.*
***Mechanical, other complication or infection and inflammatory reaction to internal orthopaedic or prosthetic device, implant or graft.*

Source: *National Center for Health Statistics, National Hospital Discharge Survey, 1988.*

Table 12: Hospitalizations Resulting from Musculoskeletal Conditions: United States, 1988 by Age and Gender*

	Rate per 10,000 population in thousands							
	All Age Groups	Less than 18	18-44	45-64	65 & over	65-74	75-84	85 & over
All Musculoskeletal Conditions	**144.2**	**50.3**	**118.3**	**187.6**	**363.9**	**283.0**	**436.8**	**621.4**
Male	**147.8**	**61.4**	**151.1**	**191.2**	**284.7**	**245.2**	**331.7**	**460.9**
Female	**140.8**	**38.7**	**86.4**	**184.2**	**418.4**	**313.1**	**500.2**	**683.4**
Fractures	**36.8**	**18.7**	**23.9**	**31.1**	**127.9**	**64.5**	**170.4**	**375.9**
Male	**35.6**	**26.0**	**33.4**	**32.5**	**75.6**	**44.4**	**99.2**	**272.8**
Female	**38.0**	**11.0**	**14.8**	**29.7**	**163.8**	**80.5**	**213.3**	**415.8**
Fracture of Neck of Femur	**10.4**	**0.6**	**0.9**	**5.2**	**71.4**	**24.7**	**98.8**	**266.3**
Male	**5.7**	**-**	**1.3**	**3.8**	**41.1**	**18.3**	**54.0**	**203.6**
Female	**14.8**	**-**	**0.6**	**6.5**	**92.2**	**29.9**	**125.8**	**290.5**
Number of Musculoskeletal Conditions (in thousands)	**3,519**	**321**	**1,229**	**863**	**1,105**	**506**	**416**	**183**

**First listed diagnosis for inpatients discharged from short-stay hospitals.*

Source: *National Center for Health Statistics, National Hospital Discharge Survey, 1988.*

hospitalized for musculoskeletal conditions. Over a ten-year period, a woman age 65 or over has, on average, a greater than 40% chance of being hospitalized for a musculoskeletal condition.

Operations

More than 3.0 million operations were performed in 1988 on the musculoskeletal system. Operations on the musculoskeletal system account for 10.6% of all inpatient procedures and were the fourth largest procedure category (Figure 6). Almost 40% of musculoskeletal procedures involve reduction of fractures (631,000) or arthroplasties (522,000) (Table 13). Arthroplasties include 251,000 joint replacements primarily involving the hip (129,000) and knee (105,000). Musculoskeletal conditions also result in significant use of ambulatory/outpatient care (National Center for Health Statistics, National Ambulatory Medical Care Survey, 1985). The most recent data available (1985) indicate musculoskeletal conditions accounted for 13.8% of all office visits and were the largest anatomic or disease category (Figure 7). Approximately 87.5 million patient visits to office-based physicians resulted from musculoskeletal conditions (Table 14).

Figure 6: All Listed Procedures for Inpatients Discharged from Short-Stay Hospitals:* United States, 1988

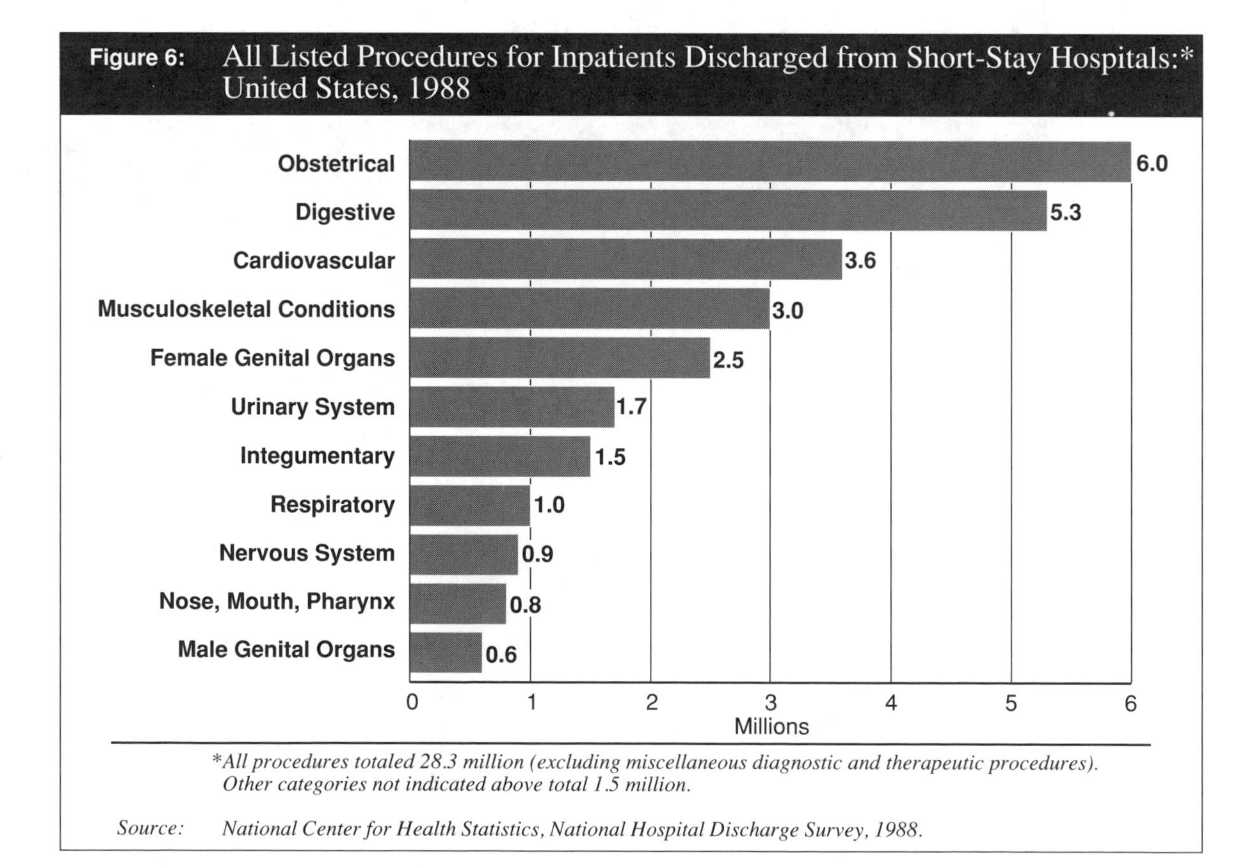

**All procedures totaled 28.3 million (excluding miscellaneous diagnostic and therapeutic procedures). Other categories not indicated above total 1.5 million.*

Source: *National Center for Health Statistics, National Hospital Discharge Survey, 1988.*

Table 13: Musculoskeletal Inpatient Procedures: United States, 1988 by Selected Aggregate Category*

Procedure		
Open Reduction of Fracture	440,000	
Other Reduction of Fracture	183,000	
Excision or Destruction of Intervertebral Disk	250,000	
Spinal Fusion	90,000	
Arthroplasty and Replacement of Knee	204,000	
Total Knee Replacement		105,000
Arthroplasty and Replacement of Hip	206,000	
Total Hip Replacement		129,000
Replacement of Head of Femur		62,000
Replacement of Acetabulum		5,000
Other Arthroplasty	112,000	
Operations on Muscles, Tendons, Fascia and Bursa	316,000	
Total, All Listed Musculoskeletal Procedures	**3,009,000**	

Excludes skull and jaw procedures.

Source: National Center for Health Statistics, National Hospital Discharge Survey, 1988.

Figure 7: Office Visits by Category of First Listed Diagnosis:* United States, 1985

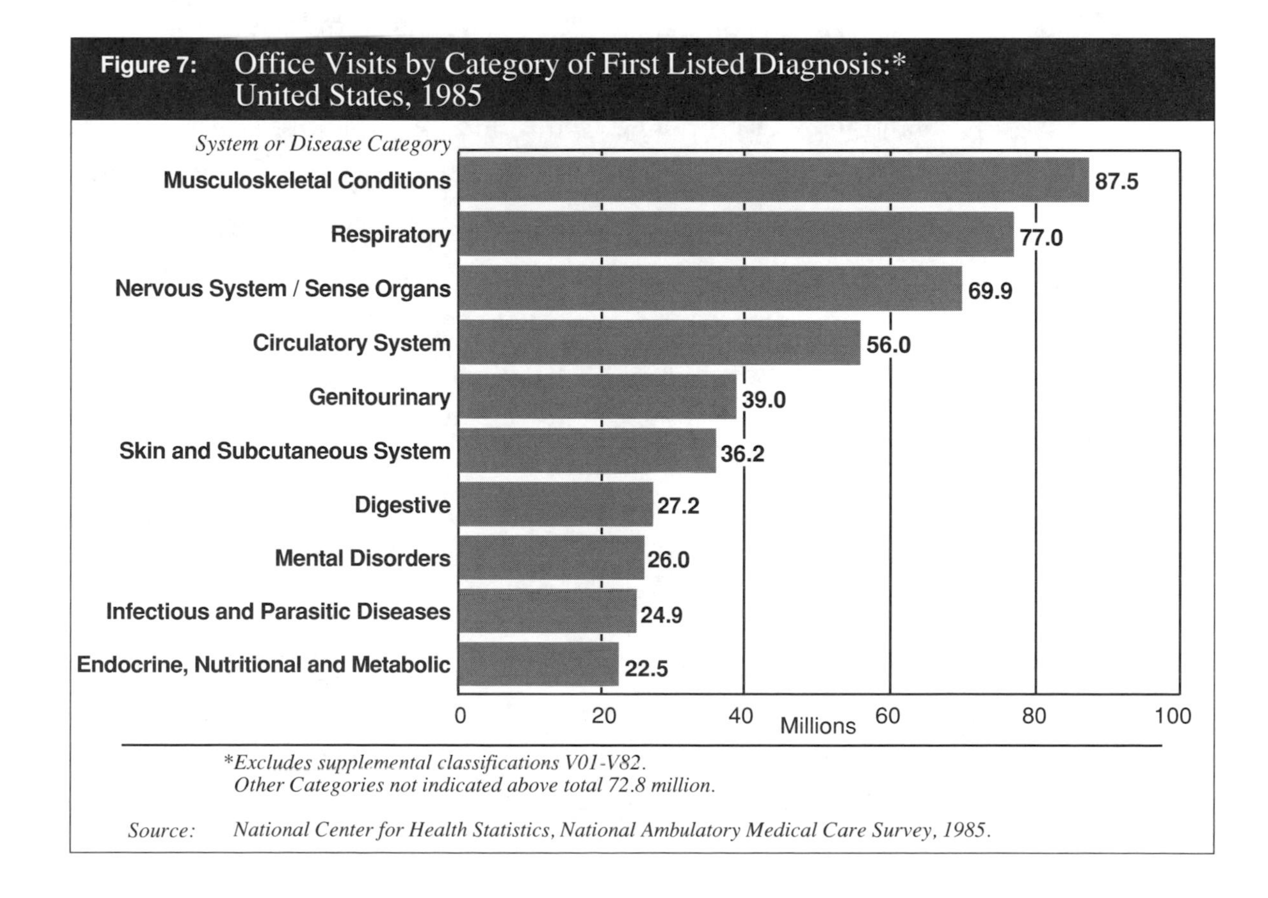

*Excludes supplemental classifications V01-V82.
Other Categories not indicated above total 72.8 million.*

Source: National Center for Health Statistics, National Ambulatory Medical Care Survey, 1985.

Table 14: Distribution of Office Visits for Musculoskeletal Conditions: United States, 1985 by Principal Diagnostic Category*

	Number of Visits (in thousands)	Percent
Musculoskeletal Diseases and Connective Tissue Disorders	**45,064**	**51.5**
Fractures	**10,128**	**11.6**
Dislocations and Sprains	**16,394**	**18.7**
Other Injuries	**10,694**	**12.2**
Complications or Reactions**	**109**	**0.1**
Congenital Anomalies	**819**	**0.9**
Neoplasms	**779**	**0.9**
Other Musculoskeletal Conditions	**3,582**	**4.1**
Total, Musculoskeletal Office Visits	**87,569**	**16.3**
Total Office Visits* **	**538,850**	**100.0**

**First listed diagnosis for inpatients discharged from short stay hospitals.*
***Mechanical, other complication or infection and inflammatory reaction to internal orthopaedic or prosthetic device, implant or graft.*
****Excludes supplemental classification V01-V82.*

Source: *National Center for Health Statistics, National Ambulatory Medical Care Survey, 1985.*

A significant source of disability, resultant health care resources, and productivity loss is the workplace. In 1988, over 6.2 million occupational injuries were reported to the Bureau of Labor Statistics for the private sector (Bureau of Labor Statistics, Occupational Injuries and Illnesses by Industry, 1988). Of these, almost half (2.9 million) resulted in lost work time or restricted work activity. An additional 116,000 cases were reported for disorders associated with repeated trauma. Repetitive trauma disorders were listed under "occupational illness." The number cited includes "conditions due to repeated motion, pressure or vibration" such as carpal tunnel syndrome.

Nationwide, there were an estimated 1.93 million severe (resulting in three or more lost work days) workplace injuries in 1987 (Rossman, et al., 1991). Of these, 590,000 (31%) required hospitalization and 11% resulted in permanent disability. Among severe injuries by type, sprains and strains accounted for 47%, fractures and dislocations 12% and severe contusions (excluding skull or head injuries) 12%. By anatomic site, lower back injuries account for 27% of severe injuries.

Musculoskeletal conditions have an enormous impact on the population of the United States. They rank highest among disease groups when indicators of the quality of life, such as impairment, disability, or limitation of activity are considered. Musculoskeletal conditions also rank first in frequency of visits to physicians, second in frequency of hospitalizations, and fourth in frequency of surgical procedures performed within hospitals.

With the aging of the United States population, the relative impact of musculoskeletal conditions will, in all probability, increase. The number of people age 65 and older is increasing two and one half times faster than the overall population. Individuals age 85 and older, who have the highest rates of musculoskeletal impairments and hospitalizations, are increasing six times faster than the population.

References

Bureau of Labor Statistics, Occupational Injuries and Illnesses in the United States by Industry, 1988. *Bulletin* 2368. August 1990.

National Center for Health Statistics, National Ambulatory Medical Care Survey, 1985 (Data Tape).

National Center for Health Statistics, National Health Interview Survey, 1988 (Data Tape).

National Center for Health Statistics, National Hospital Discharge Survey, 1988 (Data Tape).

National Center for Health Statistics, National Nursing Home Survey, 1985 (Data Tape).

Rossman SB, Miller TR, Douglass JB. "The Costs of Occupational Traumatic and Cumulative Injuries," *The Urban Institute*, March, 1991.

Section 2

Selected Musculoskeletal Conditions

chapter 1

Back, Neck and Other Joint Pain

Joint pain is a commonly reported manifestation or result of musculoskeletal disease and injury and contributes substantially to activity limitations and disability. The cumulative lifetime prevalence of self-reported back, neck or other joint pain is indicated in Table 1. Twenty-one percent of persons 25 to 74 years of age report back, neck or other joint pain. Women are more likely to report joint pain (22.4%) than are men (19.6%). The percent reporting pain varies directly with age with 15.8% of those 25-44 reporting joint pain compared with 38.4% of those age 45-64 and 40.1% in the highest age group, 65-74. Adult whites are more likely to report joint pain (21.9%) than are blacks (15.5%) or those of other racial groups (13.7%).

The frequency of joint pain varies by anatomic site. Sixteen percent report back pain, 8.2% report neck pain and 19.0% report other joint pain.

Table 1: Prevalence of Joint Pain by Site of Joint and Selected Demographic Characteristics

(Rate per 100 persons)

		Back, Neck or Other Joint Pain	Back Pain[1]	Neck Pain[2]	Other Joint Pain[3]
Total ages 25-74 years		**21.0**	**16.0**	**8.2**	**19.0**
Male		**19.6**	**16.0**	**7.0**	**16.6**
Female		**22.4**	**16.0**	**9.4**	**21.3**
Age	25-44 years	**15.8**	**12.3**	**6.6**	**12.3**
	45-64 years	**38.4**	**20.3**	**10.1**	**25.1**
	65-74 years	**40.1**	**18.2**	**9.3**	**28.1**
Race	White	**21.9**	**16.5**	**8.6**	**19.4**
	Black	**15.5**	**13.2**	**5.6**	**16.8**
	Other	**13.7**	**11.3**	**7.2**	**12.5**

[1] *Have you ever had pain in your back on most days for at least 2 weeks?*
[2] *Have you ever had pain in your neck on most days for at least 2 weeks?*
[3] *Have you had pain or aching in any joint other than the back or neck on most days for at least six weeks?*

Source: National Center for Health Statistics, NHANESII, 1976-1980.

Extrapolating these data to the 1988 population would indicate that among the adult population age 25 to 74, approximately 23 million people have experienced back pain of at least two weeks duration, 12 million have experienced neck pain of similar duration and 27 million experienced pain in other joints lasting at least six weeks.

Back pain

In the United States, back pain is one of the most frequently reported musculoskeletal problems (Cypress, 1983; Deyo and Tsui-Wu, 1987a; Deyo and Bass, 1989) and is the second most frequently reported symptom leading persons to seek care from a physician (Cypress, 1983). Back pain has been reported to be the second leading cause of work absenteeism in the United States (after upper respiratory tract complaints) and results in more lost productivity than any other medical condition (Deyo and Bass, 1989; Rowe, 1969; Salkever, 1986).

Back pain is also a significant contributor to functional disability and has been reported as the leading cause of regular limitation of activity among young adults (Deyo and Tsui-Wu, 1987b; Kelsey et al, 1979).

Prevalence data relating to back pain are limited, however. Pain is largely subjective, often occurring without identifiable physical changes and may result from a variety of diseases, structural abnormalities, or injuries.

As indicated above, data from National Health and Nutrition Examination Survey II (NHANESII) indicate a prevalence of back pain among both men and women of 16% among persons 25 to 74 years of age. There was an age differential with the highest prevalence among the 45 to 64 age group. Rates were also higher for whites (16.5%) than either blacks (13.2%) or other racial groups (11.3%). The primary site of back pain is the lower back (85.1%); 7.9% report middle back pain and 7.0% upper back pain (Figure 1). Lower back pain represented a higher percent of total back pain in men (88.4%) than in women (81.8%) (Figure 2).

Low back pain

Using NHANESII data, Deyo and Tsui-Wu examined low back pain data in greater detail. Their analysis indicates that 13.8% of persons 25 to 74 years of age have experienced low back pain of at least two weeks duration at some point in their lives (Deyo and Tsui-Wu, 1987a). Seventy five percent of this group (10.3/100 persons) reported having low back pain of this duration within the last year, and the proportion of those indicating

Figure 1: Distribution of Back Pain

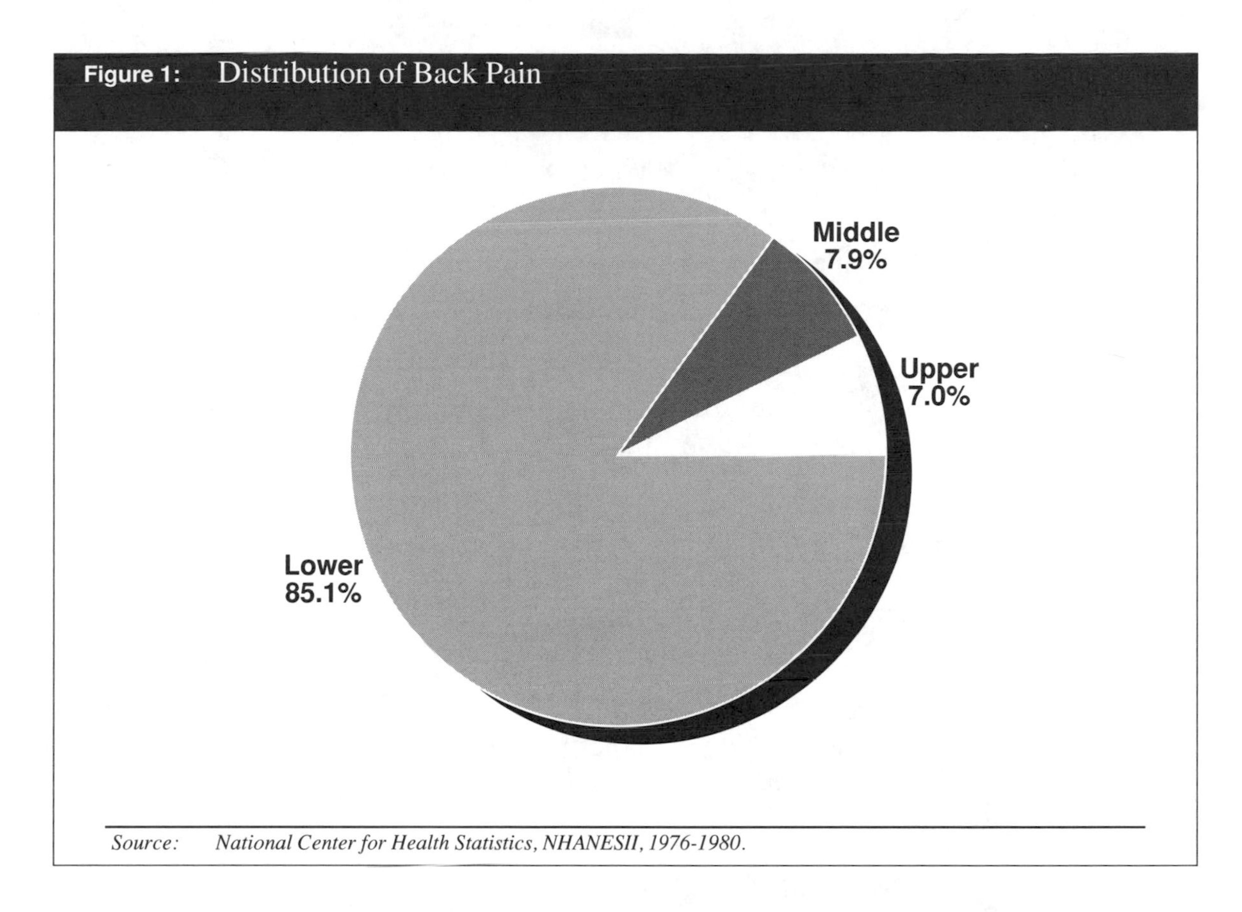

Source: *National Center for Health Statistics, NHANESII, 1976-1980.*

Figure 2: Distribution of Back Pain by Gender

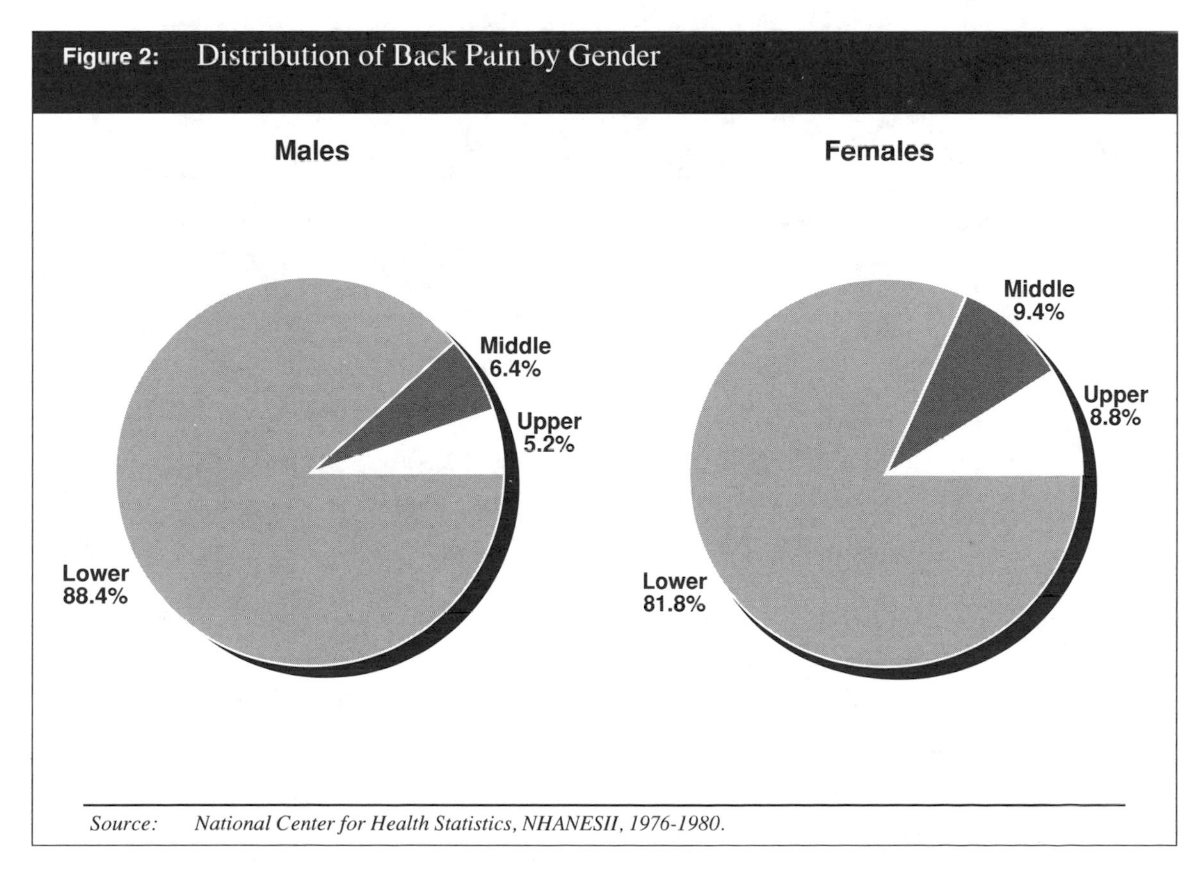

Source: *National Center for Health Statistics, NHANESII, 1976-1980.*

low back pain at any point in time (point prevalence) was 6.8% (6.8/100 persons). Back pain with features of sciatica were reported by approximately 12% (1.6/100 persons) of those with low back pain. The cumulative lifetime prevalence, one-year period prevalence and the point prevalence of low back pain all reach their maximum in the 55 to 64 year old age group. The prevalence of low back pain with evidence of sciatica peaks earlier, among those 45 to 54 years of age.

Regarding the duration and intensity of low back pain, Deyo and Tsui-Wu (1987a) indicate 33.2% report pain lasting less than one month; 33.0% report pain lasting one to five months; 32.7% report pain lasting six months or more (one percent could not remember the duration). A plurality report their low back pain was moderate (43.4%); 35.4% experienced severe pain and 21.2% mild pain.

Disk disorders

Intervertebral disk disorders are an important contributor to back pain and subsequent disablement. From the 1985 through 1988 National Health Interview Surveys, the average number of persons reporting intervertebral disk disorders annually for the period are shown in Table 2. Approximately 4.1 million persons each year report these disorders. Men were more likely to report disk disorders (2.0/100 persons) than women (1.5/100 persons). By age category, rates were highest among those in the 45 to 64 age group (3.7/100 persons).

The NHANESII survey asked persons 25 to 74 years of age if they had ever had a disk problem in their back or neck. According to Table 3, approximately 4.4 million persons reported a disk problem; (2.4 million men and 2.0 million women). Adults of working age report a greater frequency of disk problems than do the older population.

Neck pain

Neck pain is less prevalent than back pain, with 8.2% of persons in NHANESII reporting that they experienced neck pain of at least two weeks duration (Table 1). Neck pain was reported more frequently among women (9.4%) than among men (7.0%). As was the case for back pain, neck pain was reported most frequently among those 45 to 64 years of age. Rates were highest among whites (8.6%) and lowest among blacks (5.6%).

Impact

Conditions associated with back pain result in significant restrictions on activity. Data from the 1985 through 1988 National Health Interview

Table 2: Average Annual Number of Persons Reporting Intervertebral Disk Disorders: United States, 1985-1988

	Thousands	Rate per 100 persons
Total	**4,116**	**1.7**
Male	**2,275**	**2.0**
Female	**1,841**	**1.5**
Less than 18 years	***14**	**-**
18-44 years	**1,747**	**1.7**
45-64 years	**1,655**	**3.7**
65 years and over	**700**	**2.5**

** Estimate does not meet standards of reliability or precision.*

Source: *National Center for Health Statistics, National Health Interview Survey. Data Tapes, 1985-88.*

Table 3: Number of Persons 25 to 74 Years of Age Reporting Disk Problems in Back or Neck by Selected Demographic Characteristics

		(Thousands)
Total ages 25-74 years		**4,405**
Gender	Males	**2,440**
	Females	**1,965**
Age	25-44 years	**1,590**
	45-64 years	**2,189**
	65-74 years	**626**
Race	White	**4,072**
	Black	**269**
	Other	**64**

Source: *National Center for Health Statistics, NHANESII, 1976-1980.*

Surveys indicate that back or spine impairments result in an annual average of 175.8 million restricted activity days. The majority of these, 98.0 million (55.8%) occurred among women. Back and spine impairments resulted in 79.6 million bed disability days, with 38.4 million occurring among men, 41.2 million among women.

Back sprains

Among conditions associated with back pain, back sprains occur frequently and result in significant numbers of restricted activity and bed disability days. Each year between 1985 and 1988, an average of 4.6 million persons reported experiencing a back sprain (Table 4).

Table 4: Restricted Activity Days and Bed Disability Days Associated with Back Sprain: United States, 1985-88

(Thousands)

	Average Annual Incidence	Restricted Activity Days	Bed Disability Days
Total	**4,569**	**39,337**	**12,112**
Males	**2,544**	**19,535**	**4,768**
Females	**2,025**	**19,801**	**7,344**

Source: *National Center for Health Statistics, National Health Interview Survey. Data Tapes, 1985-88.*

Annually, back sprains result in over 39.3 million restricted activity days and 12.1 million bed disability days. Although the majority of back sprains were reported by men (55.7%), women accounted for the majority of restricted activity days (50.3%) and bed disability days (60.6%). Time of restricted activity per back sprain averaged 8.6 days overall and was higher among women (9.8 days) than men (7.7 days).

Over the same period, approximately 3.2 million back sprains resulted in work loss (Table 5). Sixty-two percent of back sprains associated with work-loss occurred among men. The average number of work-loss days per back sprain was 5.0 days for men and 6.3 days for women.

In addition to sprains, back conditions in general have a high prevalence in the labor force. The prevalence of deformity or orthopaedic impairments of the back among those in the labor force was reported as 72.6/1,000 persons. Prevalence was higher in females (82.2/1,000) than in males (65.1/1,000 persons (NCHS, Advance Data, No. 168).

Table 5: Work-Loss Days Associated with Back Sprain United States, 1985-88

(Thousands)

	Average Annual Incidence	Work-Loss Days	Days per Injury
Total	**3,196**	**17,537**	**5.5**
Males	**1,982**	**9,905**	**5.0**
Females	**1,214**	**7,635**	**6.3**

Source: *National Center for Health Statistics, National Health Interview Survey. Data Tapes, 1985-88.*

Back injury and compensation

Much of what is known about the impact of back pain results from data and analyses of back injury and chronic back pain in industry. Based on data from state labor departments for 1990, approximately 400,000 back injury cases resulted in disability. Back injuries accounted for 22% of cases and 31% of compensation payments (National Safety Council, Accident Facts, 1991).

Data from the Bureau of Labor Statistics' Supplemental Data System (SDS) on closed Workers' Compensation cases in nine states indicate that back injuries account for 26.2% of cases closed (Accident Facts, 1991). The average indemnity compensation for back injuries in 1990 was $5,193 and average medical payment $2,358. Among injuries affecting a single anatomic site, these payments were second only to those for neck injuries, $5,303 and $2,438 respectively. (Neck injuries accounted for 1.7% of cases closed).

Webster and Snook (1990) analyzed Liberty Mutual Insurance Company cost data from 98,999 cases of claims for low back pain initiated in 1986. The average cost per case was $6,800 with the distribution of compensation on a cost basis similar to that indicated in the SDS data above: indemnity costs, 67.2%; medical costs, 31.5%; and expense costs, 1.3%. Costs per case were highly skewed with the median cost per case being $391.

A comparison with data from a comparable study using 1980 data (Snook, 1987) indicates total compensation costs for low back pain are increasing faster than compensation costs generally. Total compensation costs for low back pain increased 241% between 1980 and 1986 compared with an increase in total Workers' Compensation costs of 184% over the same period.

Other joint pain

Joint pain located in sites other than the back or neck occurs in 19.0% of the population 25 to 74 years of age (Table 1). A greater percent of women report joint pain in other sites (21.3%) as compared with men (16.6%). Older persons are most likely to report joint pain in sites other than the back or neck, 28.1% in those 65 to 74 years of age compared with 12.3% in persons 25 to 44 years of age. There is also variation among racial groups: 19.4% of whites report joint pain in sites other than the back or neck as compared with 16.8% of blacks and 12.5% percent of other races.

The distribution of other joint pain in persons 25 to 74 years of age is shown in Table 6. Knee pain is reported more frequently than pain in other sites (45.9% of those with other joint pain). Knee pain is followed in frequency by shoulder (34.6%) and finger pain (32.4%).

Table 6: Persons 25 to 74 Years of Age Reporting Joint Pain at Sites Other than the Back or Neck, Percent by Site

Site	Percent
Knee	**45.9**
Shoulder	**34.6**
Fingers	**32.4**
Elbow	**24.4**
Hip	**22.8**
Ankle	**19.1**
Wrist	**18.0**
Foot	**15.6**

Note: *A person could report more than one site of joint pain.*

Source: *National Center for Health Statistics, NHANESII, 1976-1980.*

The percent distribution of reported joint pain (other than the back or neck) is indicated in Figure 3. Almost one-third of reported joint pain (32.3%) occurs in the knee, with 22.2% occurring in the shoulder. Among those who report pain in joints other than the back or neck, knee pain accounts for virtually the same percent among men and women. In this group, men are more likely to report shoulder, elbow and foot pain; women are more likely to report finger, hip, ankle and wrist pain (Figure 4).

Figure 3: Percent Distribution of Joint Pain by Site of Joint Pain

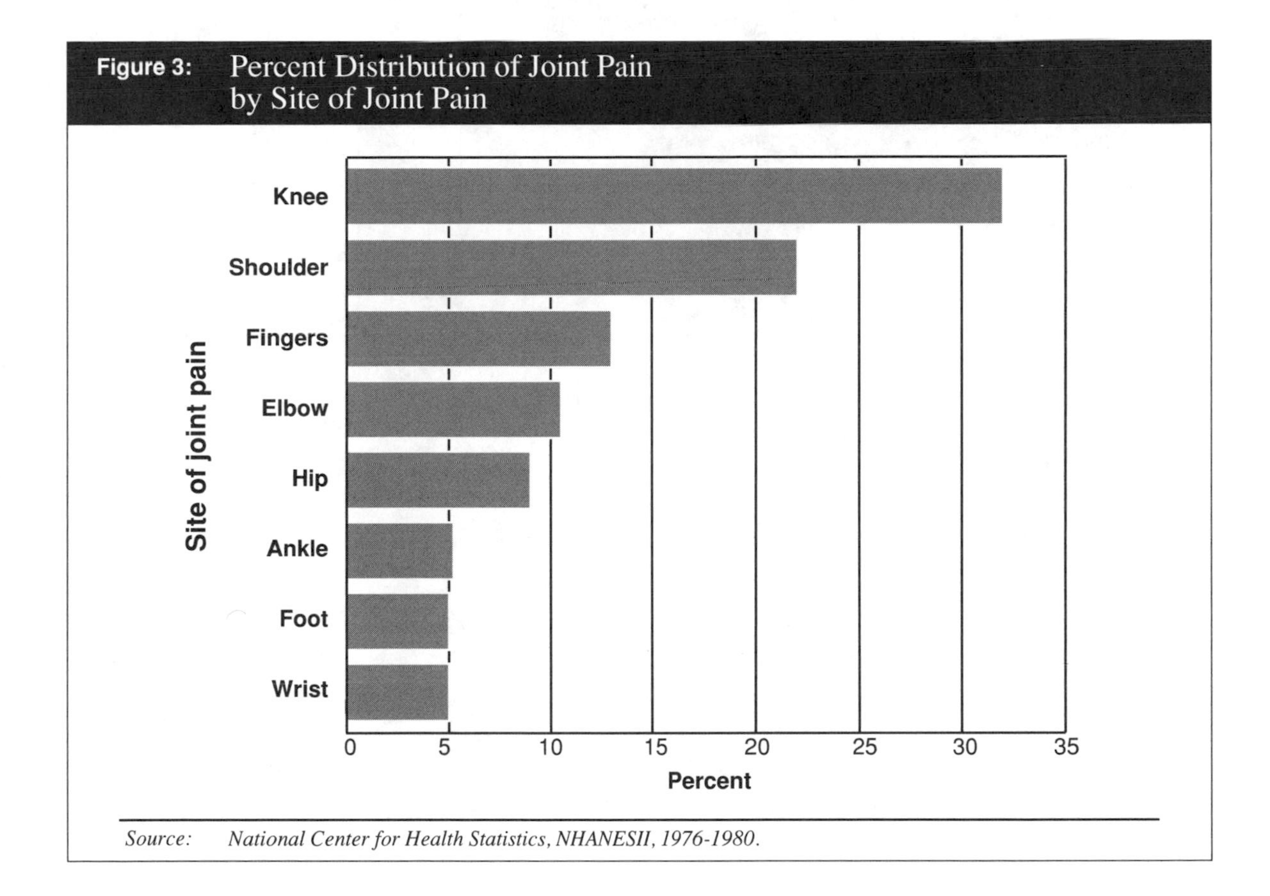

Source: *National Center for Health Statistics, NHANESII, 1976-1980.*

Figure 4: Percent Distribution of Joint Pain, by Site of Joint Pain, by Gender

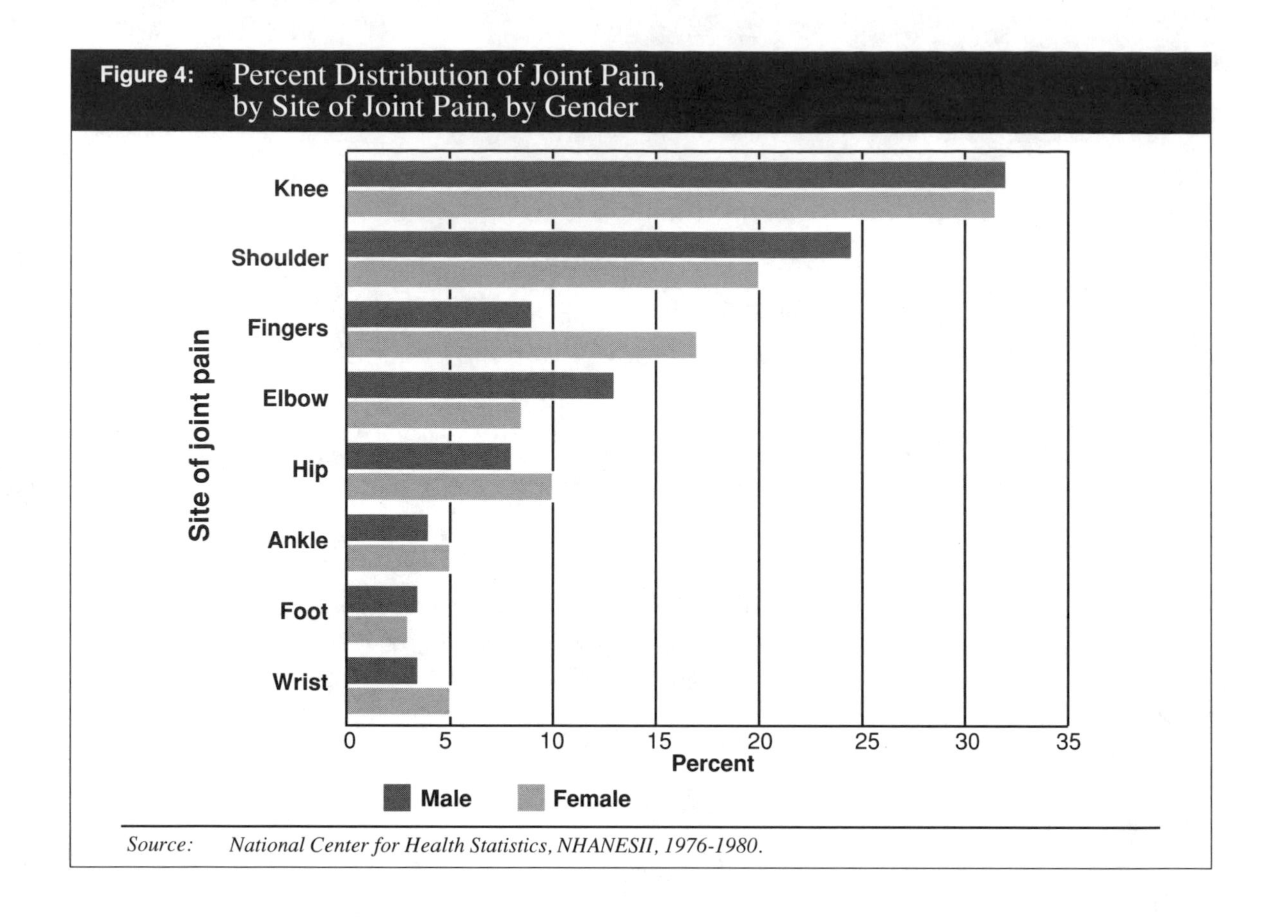

Source: *National Center for Health Statistics, NHANESII, 1976-1980.*

References

Cypress BK. Characteristics of physician visits for back symptoms: A national perspective. *Am J Public Health*, 73:389-395, 1983.

Deyo RA, Tsui-Wu YJ. Descriptive Epidemiology of Low-Back Pain and its Related Medical Care in the United States. *Spine*, Vol. 12(3), 1987a.

Deyo RA, Tsui-Wu YJ. Functional Disability Due to Back Pain. A population-based study indicating the importance of socioeconomic factors. *Arthritis and Rheumatism*, Vol. 30 (11), 1987b.

Deyo RA, Bass JE. Lifestyle and low-back pain. The influence of smoking and obesity. *Spine*, Vol. 14(5), 1989.

Kelsey JL, White AA, Pastides H, Bisbee GE. The Impact of Musculoskeletal Disorders on the Population of the United States. *J Bone Joint Surg*, 61A:959-964, 1979.

Miles T. Glegal K, Harris T. Musculoskeletal Disorders - Time Trends, Co-Morbid Conditions. In: *Health Data on Older Americans*: 1991, National Center for Health Statistics, forthcoming 1992.

National Center for Health Statistics. Advance data, No. 168, *Vital and Health Statistics.*

National Center for Health Statistics. National Health Interview Survey, 1985-1988 (Data tapes).

National Center for Health Statistics. Second National Health and Nutrition Examination Survey, (NHANESII) 1976-1980.

National Safety Council. *Accident Facts*, 1991 Chicago.

Rowe ML. Low Back Pain in Industry. A Position Paper. *J Occup Med,* 11:161-169, 1969.

Salkever DS. Morbidity costs: National Estimates and Economic Determinants. NCHSR Research Summary Series, October, 1985, Department of Health and Human Services Publication No. (PHS) 86-3393, 1986.

Snook SH. The Costs of Back Pain in Industry. *Spine: State of the Art Reviews*, Vol. 2, Sept. 1987.

Webster BS, Snook SH. The Cost of Compensable Low Back Pain, *J Occup Med*, VOl. 32, Jan. 1990.

chapter 2

Arthritis

The various forms of arthritis affect persons of all ages and have a substantial impact on activity levels and the quality of life. Although there are many types of arthritis, by far the most common types in Western populations are osteoarthritis and rheumatoid arthritis. Other less common but nevertheless important forms of arthritis include gout, ankylosing spondylitis, and juvenile rheumatoid arthritis.

Prevalence

The average annual prevalence rates for arthritis and selected rheumatic diseases by gender and age for 1985 through 1988 are indicated in Table 1. Averaging the national data over the four-year time period allows for statistical stability of the estimates. These data are based on personal interviews and indicate a prevalence rate of 130.3/1,000 persons for arthritis with an overall male-female prevalence ratio of 1:1.8. Prevalence varies directly with age and remains higher for women in all age categories (Figure 1).

Table 1: Average Annual Prevalence Rates of Selected Rheumatic Diseases by Gender and Age: United States, 1985-88*

(Rate per 1,000 persons)

	Arthritis	Nonarticular Rheumatism	Bursitis	Gout
Total	**130.3**	**2.0**	**19.0**	**9.3**
Gender				
Male	**91.9**	**1.9**	**15.7**	**13.3**
Female	**166.2**	**2.2**	**22.0**	**5.6**
Age				
0-17 years	**2.4**	****0.04**	**0.9**	**-**
18-44 years	**51.5**	**1.3**	**15.4**	**2.9**
45-64 years	**270.8**	**3.7**	**41.0**	**22.4**
65 years & over	**480.4**	**6.7**	**37.2**	**32.8**

**Self-reported cases*
***Estimate does not meet standards of reliability or precision.*

Source: *National Center for Health Statistics, National Health Interview Survey, 1985-88.*

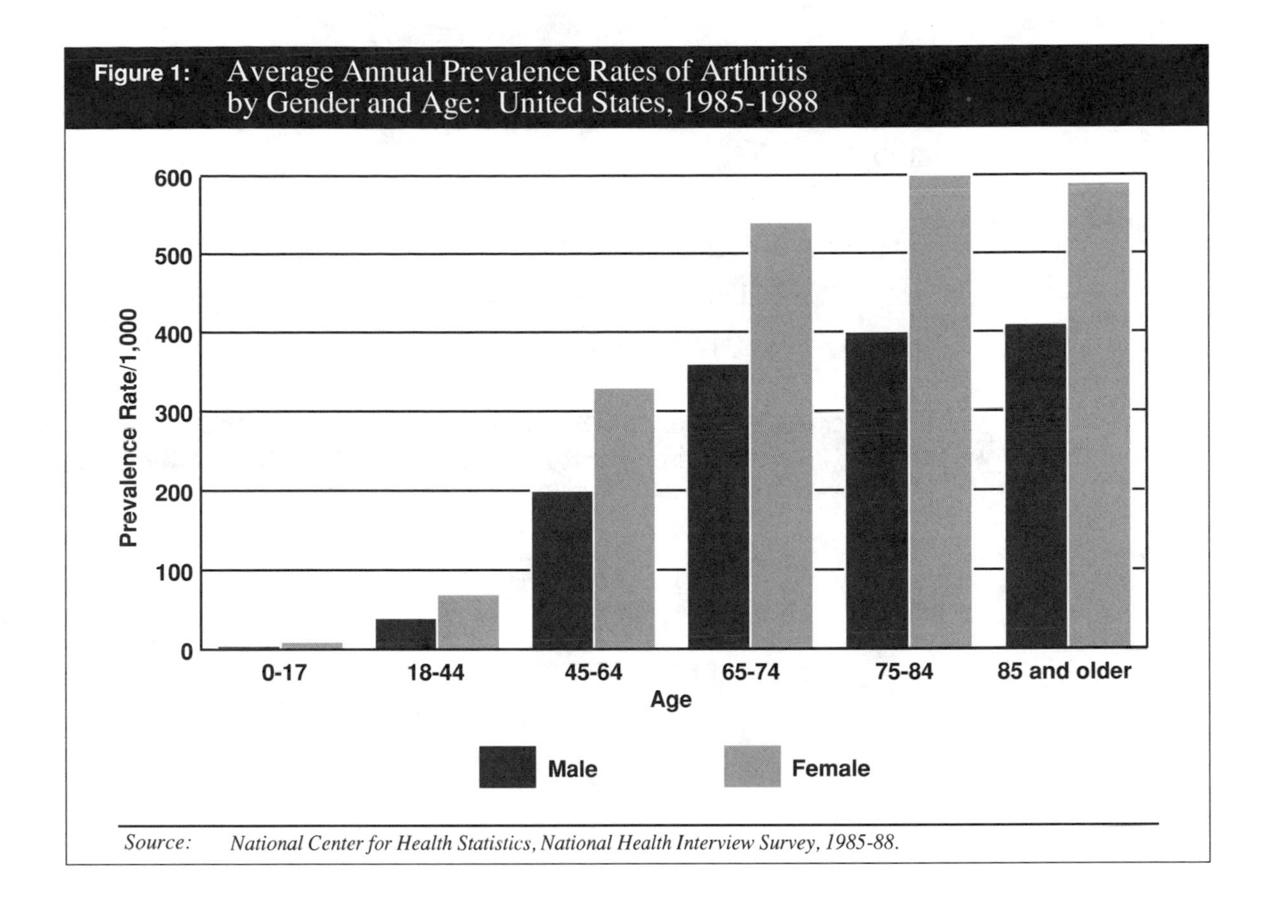

The other rheumatic diseases show substantially less prevalence. All four diseases, however, show prevalence increasing with age and, with the exception of gout, are more prevalent among women.

Recent national survey data indicate that arthritis is the second most frequently reported chronic condition (30.3 million conditions) by the civilian non-institutionalized population of the United States (NCHS, Advance Data, No. 155). Among females, arthritis ranks as the most frequently reported condition. Among males, arthritis ranks fifth in prevalence following chronic sinusitis, deformities or orthopaedic impairments, hypertension and hearing impairments.

Arthritis is reported by nearly half of those persons 65 years of age and older and is present in more than half of the women 65 years of age and older, with black women having greater prevalence than white women in some reports (Health Statistics of Older Persons, 1986) and having about the same prevalence according to other reports (Cunningham and Kelsey, 1984). Although exact prevalence estimates of arthritis depend on the method of diagnosis, arthritis, primarily osteoarthritis, consistently is the most frequently reported chronic condition of those 65 years of age and older.

Within the 65 and older age group, prevalence continues to increase as age increases (Longitudinal Study on Aging, NCHS, 1984) Prevalence for the population 80 years of age and older living in the community reaches 557.3/1,000 (Table 2).

Table 2: Prevalence Rates for Arthritis Among Community Living Persons Age 70 and Over

	(Rate per 1,000 persons)
Total, 70 years and over	**549.4**
Male	**452.2**
Female	**610.6**
70-79 years	**546.4**
80 years and over	**557.3**

Source: *The Longitudinal Study on Aging, NCHS, 1984.*

Table 3 shows the prevalence estimates for persons 55 years of age and older who reported a history of physician-diagnosed arthritis in one of three national surveys conducted over the time period between 1960 and 1984 (Health Statistics of Older Persons Vol. II, forthcoming). Between 37% and 48% of persons 55 years of age and older reported physician-diagnosed arthritis during this time period. In general, this percent increased with age and was higher for women than men. White and black men were equally as likely to report physician-diagnosed arthritis in the first two time periods, but in the latest survey, 48% of black men reported this condition as compared with 38% of white men. White and black women were equally as likely to report diagnosed arthritis in the first time period, but in the two other time periods black women indicated greater prevalence of this condition than did white women.

Table 3: Percent of Persons 55 Years of Age & Over Who Reported Physician-Diagnosed Arthritis: United States, Selected Time Periods by Gender and Race

Gender, Race, and Age	1960-62	1976-80	1984
Total	37.2	44.3	47.7
55-59 years	34.7	38.9	38.4
60-64 years	37.0	44.2	44.6
65-69 years	35.9	48.4	48.8
70-74 years	42.1	48.0	55.5
75-79 years	39.7	-	54.2
80 years & over	-	-	55.6
Male	28.3	37.2	38.9
55-59 years	24.4	32.6	29.2
60-64 years	28.8	38.2	37.0
65-69 years	32.1	40.9	42.2
70-74 years	28.4	38.7	47.8
75-79 years	29.7	-	48.3
80 years & over	-	-	45.1
White Male	28.1	37.3	38.2
55-59 years	22.8	33.5	28.2
60-64 years	28.8	38.0	36.1
65-69 years	32.0	40.3	41.3
70-74 years	29.2	38.9	47.4
75-79 years	30.5	-	42.4
80 years & over	-	-	44.8
Black Male	30.2	36.4	47.5
55-64 years	33.6	31.9	43.3
65-74 years	26.8	42.8	51.7
75 years & over	*18.8	-	51.1
Female	45.1	50.2	54.0
55-59 years	43.7	44.1	45.5
60-64 years	45.1	50.0	50.9
65-69 years	39.0	54.2	54.0
70-74 years	53.3	54.9	60.7
75-79 years	49.8	-	61.3
80 years & over	-	-	60.4
White Female	45.1	49.2	53.2
55-59 years	43.1	42.2	45.5
60-64 years	44.5	49.6	50.4
65-69 years	39.4	52.8	52.9
70-74 years	53.1	55.2	59.3
75-79 years	53.0	-	60.2
80 years & over	-	-	59.3
Black Female	45.0	59.7	61.2
55-64 years	49.9	58.5	50.1
65-74 years	44.5	61.3	69.9
75 years & over	*19.7	-	73.6

Question: Did a doctor confirm or ever tell you that you had arthritis?
*Estimate does not meet standards of reliability or precision.

Sources: *National Center for Health Statistics: 1960-62 data from the Health Examination Survey; 1976-80 data from the second National Health and Nutrition Examination Survey; 1984 data from the National Health Interview Survey Supplement on Aging.*

Impact

The influence of arthritis on the health status and activity levels of the elderly, on a cross-sectional basis, has been well established (Yelin and Katz, 1990). When examined by physicians in community-based studies of random samples of the population, more than half of the elderly meet strict diagnostic criteria for arthritis, even though many persons diagnosed in epidemiologic studies neither report symptoms nor report having received a diagnosis of arthritis from their physicians (Yelin et al, 1983). Nevertheless, at least one-third of the men and one-half of the women 65 years of age and older report arthritis, and between 8% and 13% report some level of activity limitation as a result. Arthritis has been reported to be the most common cause of activity limitation among elderly women and is the second leading cause among elderly men (Verbrugge, 1984; LaPlante, 1988).

Arthritis is a major contributor to functional impairment among the older population. It is a major cause of disability and accounts for a large proportion of the hospitalizations and health care expenditures of the elderly (Cunningham and Kelsey, 1984). Among the chronic conditions of the elderly, arthritis and rheumatism rank first when measured as the percentage of persons limited in activity (Epstein et al, 1986). Approximately five million persons are limited in some way by arthritis, with 20% to 40% of these unable to perform their major activity.

Assessing the impact of arthritis is complicated somewhat by the existence of co-morbid conditions. Of persons with arthritis who are limited to a certain extent in their activities, 56% claim that the principal cause of their limitation is another chronic condition (Epstein et al, 1986). The addition of one or more non-arthritic conditions substantially increases the probability of limitation of some activity. Arthritis, in turn, is the major cause of limitation in 22% of those individuals who also report another chronic condition.

Arthritis is estimated to be the principal cause of total incapacitation for about one million persons aged 55 and older and is thought to be an important contributing factor for about one million more (Epstein et al, 1986). Approximately 10% of elderly arthritics visit their physician monthly for this condition, twice as often as other elderly persons do for any other condition (Epstein et al, 1986). For those persons aged 65 and older, it has been reported that 7% of doctor visits, 3% of hospital days, and 17% of bed days are due to arthritis and rheumatism (Guralnik and Fitzsimmons, 1986).

A recent longitudinal analysis assessed the impact of arthritis on the functional ability of older persons (Yelin and Katz, 1990). Fifty-six percent of the 70 and older population reported arthritis, 14% had arthritis by itself and 41% had arthritis along with other chronic conditions that might affect functional ability. Twenty-seven percent had other conditions and no arthritis and 18% had none of the selected conditions. The study indicated a greater percent of persons with arthritis were disabled two years after the baseline survey. Among persons with only arthritis over the two-year period, the percentage reporting no physical limitations declined from 33% to 22%. Among persons with arthritis and other conditions, only 35% of those with no limitations at baseline were disability-free two years later.

Osteoarthritis

There are several types of arthritis, with osteoarthritis and rheumatoid arthritis being predominant. Osteoarthritis, a slowly progressive disease characterized by degeneration of articular cartilage, with proliferation and remodeling of subchondral bone, is the most common type (Lawrence et al, 1989; Roach, 1991). It presents with a clinical picture of pain, deformity, and loss of joint motion (Mankin, 1989). The prevalence of osteoarthritis increases with age and the mean age of onset varies by joint (Acheson, 1975; Peyron, 1986; Roach, 1991).

Estimating the prevalence of osteoarthritis is difficult for several reasons (Lawrence et al, 1989). First, osteoarthritis is diagnosed objectively on the basis of reading radiographs, but many people with radiographic evidence of this condition have no symptoms. In estimating prevalence levels, it is debatable whether these individuals should be considered as diseased. Second, the prevalence of osteoarthritis is based on clinical evaluation because radiologic data are not available for all joints of the body. Third, prevalence varies between studies depending on whether only moderate and severe radiologic changes are counted or whether mild changes are also included. Finally, there is a lack of data on the oldest old (85 years of age and older), the group with the greatest prevalence.

In light of these difficulties, few estimates of the prevalence of osteoarthritis are available. Lawrence et al (1989) reported on three surveys, which estimated this prevalence in different ways. One survey, the Tecumseh Community Health Study, used clinical evaluation and medical history diagnosis or physical examination diagnosis. Both methods indicate that the prevalence of this disease is greater in women than men and increases with age. Of note, however, is the difference in prevalence for the older population depending on the type of diagnosis (Table 4). In the 60 and older age group, prevalence levels are 19% higher for men and 38% higher for women when the diagnosis is based on physical examination.

Table 4: Prevalence of Osteoarthritis as Diagnosed by History or Examination by Gender and Age Group

(Rate per 100 persons)

	Diagnosis by History		Diagnosis by Examination	
Age	Males	Females	Males	Females
Less than 20 years	-	-	-	-
20-39 years	**0.2**	**0.4**	-	**0.2**
40-59 years	**3.4**	**8.4**	**4.0**	**8.9**
Over 60 years	**17.0**	**29.6**	**20.3**	**40.8**
All ages	**1.9**	**4.0**	**2.2**	**5.0**
All ages over 20 years	**4.5**	**7.3**	**4.2**	**9.0**

Source: *Lawrence et al, 1989 (Mikkelsen).*

In the NHANESI survey, the data on osteoarthritis were based on a thorough musculoskeletal evaluation during which were noted any abnormalities of the spine, knees, hips and peripheral joints as well as any specific diagnoses. Based on these examinations, 12% of the US population 25 to 74 years of age have osteoarthritis (Lawrence et al, 1989; Cunningham and Kelsey, 1984).

Results of the third survey, in which osteoarthritis is defined on the basis of radiologic changes in specific joints regardless of physical symptoms, are shown in Table 5. Of note is the increasing prevalence with increasing age for each joint. Also of interest is the crossover in gender-specific rates at varying ages. In the younger age groups, prevalence is greater among men than women, but this situation is reversed as the population ages.

Table 5: Prevalence of Radiologic Changes Indicative of Osteoarthritis by Age and Gender

(Rate per 100 population)

			Mild, Moderate and Severe			*Moderate and Severe*		
Site	Ages	Source	Males	Females	Total	Males	Females	Total
Hands	18-79	NHES	29.4	30.4	29.9	5.3	9.9	7.1
	25-74		32.0	33.0	32.5	5.4	0.2	7.9
	18-24		2.8	0.4	1.6	-	-	-
	25-34		4.8	2.1	3.4	0.1	-	-
	35-44		17.5	11.3	14.3	0.6	1.1	0.9
	45-54		39.0	34.0	36.4	1.8	5.5	3.7
	55-64		56.6	68.8	63.0	12.6	21.5	17.3
	65-74		71.0	77.1	74.5	22.4	37.0	30.7
	75-79		78.7	88.4	84.5	33.2	51.0	43.9
Feet	18-79	NHES	19.8	21.3	20.6	1.5	2.9	2.2
	25-74		21.1	23.2	22.2	1.6	3.0	2.3
	18-24		4.5	1.2	2.8	-	-	-
	25-34		9.7	4.4	7.0	-	-	-
	35-44		17.3	11.2	14.1	0.4	0.4	0.4
	45-54		22.8	25.0	23.9	1.5	1.9	1.7
	55-64		29.0	44.1	36.9	3.4	6.9	5.2
	65-74		40.3	47.1	44.2	5.8	9.1	7.7
	75-79		48.6	53.1	51.3	4.8	14.6	10.7
Knees	25-74	NHANESI	2.6	4.9	3.8	0.5	1.3	0.9
	25-34		-	0.1	0.0	-	0.0	0.0
	35-44		1.7	1.5	1.6	0.1	0.5	0.3
	45-54		2.3	3.6	3.0	0.2	0.5	0.4
	55-64		4.1	7.3	5.7	1.0	0.9	0.9
	65-74		8.3	18.0	13.8	2.0	6.6	4.6
Hips	25-74	NHANESI	1.3	-	-	0.5	-	-
	25-34		0.4	-	-	0.2	-	-
	35-44		0.1	-	-	-	-	-
	45-54		0.7	-	-	0.1	-	-
	50-54		-	0.8	-	-	0.1	-
	55-64		2.6	2.8	2.7	0.7	1.6	1.2
	65-74		4.6	2.7	3.5	2.3	1.2	1.7
	55-74		3.5	2.8	3.1	1.4	1.4	1.4

Sources: *Lawrence et al, 1989.*
NHES source Engel, 1966.
NHANESI source Maurer, 1979.

Rheumatoid arthritis

Rheumatoid arthritis is a chronic systemic inflammatory disease of unknown etiology. It is characterized pathologically by a destructive synovitis in multiple diarthrodial joints. Autoantibodies to immunoglobulin IgG (rheumatoid factor RF) in both the blood and joints can be demonstrated in the majority of patients (Smith and Arnett, 1991). Rheumatoid arthritis may be mild and relapsing or severe and progressive, leading to joint deformity and incapacitation. Typically, the small joints of the hands and feet are involved early, but virtually any synovial joint, including the large weight-bearing joints, may eventually be affected (Smith and Arnett, 1991).

There is apparently an important genetic component to the etiology of rheumatoid arthritis. The strongest evidence comes from the association of the disease with human leukocyte antigen (HLA-DR4), a genetically determined allele of the major histo-compatibility complex located on the short arm of chromosome six (Bodmer, 1987).

It is generally agreed that rheumatoid arthritis has a worldwide distribution and affects all racial and ethnic groups. The exact prevalence in the United States is unknown but has been estimated to range between 0.5% and 1.5% (Smith and Arnett, 1991).

Rheumatoid arthritis occurs at all age levels and generally increases in prevalence with advancing age. It is two to three times more common in women than in men and peaks in incidence between the fourth and sixth decades (Kelly, 1985). In addition to immunological factors, environmental, occupational, and psychosocial factors have been studied for their possible etiologic role in the development of rheumatoid arthritis.

Prevalence surveys of rheumatoid arthritis have for the most part relied on two sets of criteria for case definition: the American Rheumatism Association (ARA) criteria and the New York criteria for rheumatoid arthritis (Lawrence et al, 1989). Two independent studies noted that the majority of persons classified as having probable rheumatoid arthritis in population surveys do not have clinical rheumatoid arthritis (Lawrence, 1977; O'Sullivan and Cathgart, 1972).

The National Health Examination Survey is the largest population survey of rheumatoid arthritis in the United States. Based on ARA criteria, 0.9% of the population (0.7% of men and 1.6% of women) were classified as having definite rheumatoid arthritis. The prevalence increased with increasing age: 2% of men 55 and older, 3% of women 55 to 64 years of age, and 5% of women 65 years of age and older. There was no difference in prevalence between blacks and whites.

In a community-based study of rheumatoid arthritis in Sudbury, Mass., definite rheumatoid arthritis was identified in 0.9% of respondents 15 to 75 years of age (Cathgart and O'Sullivan, 1970). A male:female prevalence ratio of 1:3 was found, along with an increase in prevalence with increasing age.

Cunningham and Kelsey (1984) reported on the prevalence of rheumatoid arthritis in NHANESI based on secondary data analysis. The case definition was based on physician examination, not radiographic data. The prevalence rate was 0.87%, and increased with increasing age. The male to female ratio was 1:2.

Although most population surveys do not collect incidence data on rheumatoid arthritis, it is suspected that the incidence rate is relatively stable (Smith and Arnett, 1991). A decline in incidence in women, however, has been noted recently with a possible link to the use of oral contraceptives and postmenopausal estrogens. In a study by Linos et al (1980) in Rochester, Minnesota, an average annual incidence rate of 28.1/100,000 was noted for men; for women, the rate was 65.7/100,000. The secular trend differed by gender, with male rates remaining stable and the female rates declining.

References

Acheson RM, Collart AB. New Haven Survey of Joint Diseases: XVII. Relationship Between Some Systemic Characteristics and Osteoarthrosis in a General Population. *Ann Rheum Dis*, 34:379-387, 1975.

Bodmer WF. HLA, 1987. In Dupont, B (ed.): *Immunobiology of HLA, Vol. 2, Immunogenetics and Histocompatibility*. New York, Springer Publishing Company, 1-9, 1989.

Cathgart ES, O'Sullivan JB. Rheumatoid Arthritis in a New England Town: a Prevalence Study in Sudbury, Massachusetts. *N Engl J Med*, 282:421-424, 1970

Cunningham LS, Kelsey JL. Epidemiology of Musculoskeletal Impairments and Associated Disability. *Am J Public Health*, 74:574-579, 1984.

Engel A. Osteoarthritis in Adults by Selected Demographic Characteristics, United States, 1979. Vital and Health Statistics Series 10, No. 136, USDHHS, 1981.

Epstein WV, Yelin EH, Nevitt M. Kramer JS. Arthritis: A Major Health Problem of the Elderly. In: Moskowitz, RW, Hang, MR, eds. *Arthritis and the Elderly*. New York, Springer Publishing Company, 1986.

Guralnik JM, Fitzsimmons SC. Aging in America: A Demographic Perspective. *Geriatric Cardiology*, 4:175-183, 1986.

Health Statistics of Older Persons, United States, 1986. (1987) *DHHS Publication No. (PHS) 87-1409*. U.S. Department of Health and Human Services, Public Health Service, NCHS, Hyattsville, MD.

Health Statistics of Older Persons, Volume II, United States, 1991. Forthcoming from Department of Health and Human Services, Public Health Service, NCHS, Hyattsville, MD.

Kelley WN, Harris ED, Ruddy S, Sledge CB. *Textbook of Rheumatology, ed. 2*, Philadelphia, W.B. Saunders Co., 1985.

Kelsey JL. *Epidemiology of Musculoskeletal Disorders*. New York: Oxford University Press, 1982.

LaPlante M. *Data on Disability from the National Health Interview Survey, 1983-1985*. Corte Madera, California, InfoUse, Inc. 1988.

Lawrence RC, Hoohberg MC, Kelsey JL, McDuffie FC, Medsger TA, Felts WR, Shulman LE. Estimates of the Prevalence of Selected Arthritic and Musculoskeletal Diseases in the United States. *J Rheumatology*, 16:4, 427-441, 1989.

Lawrence JS. *Rheumatism in Populations*, London: Wm. Heinemann Medical, 156-271, 1977.

Linos A, Worthington JW, O'Fallon WM, and Kurland LTG: The Epidemiology of Rheumatoid Arthritis in Rochester, Minnesota. A Study of Incidencc, Prevalence, and Mortality. *Am J Epidemiology*, 11:87, 1980.

Mikkelsen WM, Dodge HJ, Duff IF, et al. Estimates of the Prevalence of Rheumatic Diseases in the Population of Tecumseh, Michigan, 1959-1960. *J Chronic Dis*, 20:351-369, 1967.

NCHS, Advance Data, No. 155, Vital and Health Statistics.

O'Sullivan JB, Cathgart ES. The Prevalence of Rheumatoid Arthritis: Followup Examination of the Effect of Criteria on Rates in Sudbury, Massachusetts. *Ann Intern Med*, 76:572-577, 1972.

Peyron JG. Osteoarthritis: The Epidemiologic View Point. *Clin Orthop*, 213:13-19, 1986.

Roach K. The Association Between the Biomechanical Aspects of Past Occupation and Osteoarthritis of the Hip in Men. Unpublished doctoral dissertation, University of Illinois, 1991.

Smith CA, Arnett FC. Epidemiologic Aspects of Rheumatoid Arthritis, Current Immunogenetic Approach. *Clin Orthop*, No. 265, 23-35, 1991.

Verbrugge L. Longer Life but Worsening Health? Trends in Health and Mortality of Middle-aged and Older Persons. *Milbank Q* 62:475-519, 1984.

Yelin EH, Kramer J, Epstein W. Arthritis Policy and the Elderly, Policy Paper, No. 5. University of California, San Francisco, Aging Health Policy Center, 1983.

Yelin EH, Katz PP. Transitions in Health Status Among Community-dwelling Elderly People with Arthritis. *Arth and Rheum*, Vol. 33, No. 8, August, 1990, 1205-1215.

chapter 3

Osteoporosis

The epidemiology of osteoporosis, a disease manifested by diminished bone mineral mass and highly associated with an increased susceptibility to fracture, is not well documented (Cummings et al, 1985). The measurement of the prevalence of osteoporosis is difficult because

1. some progressive loss of bone mass appears to be a function of the aging process
2. bone loss can be caused by other disease processes
3. the bone mass at one site is not highly correlated with the bone mass at other sites
4. only in the last decade, with the combination of absorptiometry and CT scanning techniques, has measurement of bone mass been made reliable (Cummings et al, 1985; Riggs et al, 1981; Mazess, 1982).

Prevalence

A seminal study that used radiologic evidence to determine the age-specific prevalence of osteoporosis in women was conducted in Michigan (Iskrant and Smith, 1969). As indicated in Table 1, the prevalence of radiologically defined osteoporosis is directly related to age and increases from 18% for women 45 to 49 years of age to 89% for those 75 years of age and older.

There are no national survey data using objective diagnostic criteria for osteoporosis. Self-reported data, however, from the Longitudinal Study on Aging from the National Center for Health Statistics is indicated in Table 2. As these data indicate, the use of self-reported data, in all likelihood, underestimates the true prevalence of this condition because these rates are considerably lower than those obtained using radiologic evidence. Underreporting is likely to result because osteoporosis may not be diagnosed until a fracture occurs. Nevertheless, the trend of increasing prevalence with increasing age is evident, as is the higher rate in women than in men.

Fractures

The presence of osteoporosis predisposes one to fractures of the hip, vertebrae, distal forearm, humerus and pelvis.Fractures at these sites that result from "minimal trauma" are considered osteoporotic (Cummings et al, 1985).

Table 1: Prevalence of Radiologic Evidence of Osteoporosis in the Dorsolumbar Spine in Michigan Women of Age 45 Years and Older, by Age

Age	*Rate per 100 persons**
45-49 years	**17.9**
50-54 years	**39.2**
55-59 years	**57.7**
60-64 years	**65.5**
65-69 years	**73.5**
70-74 years	**84.2**
75 years & older	**89.0**
Total	**56.7**

**Moderate and severe osteoporosis seen on radiographs*

Source: *Iskrant and Smith, 1969.*

Table 2: Prevalence of Self-Reported Osteoporosis and Hip Fracture in Persons 70 Years of Age and Older: United States, 1984

	(Rate per 100 persons)		
Type of Impairment	Osteoporosis	Hip Fracture	Either Condition
Total (70 years & older)	**3.7**	**4.5**	**7.7**
70-79 years	**3.5**	**3.2**	**6.2**
80 years & older	**4.3**	**8.0**	**11.6**
Male	**0.6**	**2.7**	**3.3**
70-79 years	**0.7**	**1.9**	**2.5**
80 years & older	**0.4**	**5.5**	**5.9**
Female	**5.7**	**5.6**	**10.5**
70-79 years	**5.5**	**4.0**	**8.8**
80 years & older	**6.2**	**9.2**	**14.4**

Source: *National Center for Health Statistics, Longitudinal Study on Aging.*

Fractures of the hip are a major consequence of osteoporosis in terms of disability, medical costs, and mortality (Lawrence et al, 1989; Cooper, 1989; Cummings et al, 1985, Lewinnek et al, 1980). About 250,000 hip fractures occur annually in the United States (Lawrence et al, 1989; Graham, 1988). They are discussed in greater detail later in this section.

There have been few epidemiologic studies of vertebral fracture primarily because these fractures are usually asymptomatic at onset and are difficult to define radiographically (Cooper, 1989; Cummings et al, 1985). Vertebral fractures may be partial or complete compression fractures. A study in Denmark found that 4.5% of 70-year-old women had crush fractures and an additional 18% had partial vertebral fractures (Jensen et al, 1982). Another study conducted on female hospital outpatients in the United Kingdom found crush fracture prevalence to be 2.5% for those 60 years of age and 7.5% for those 80 years of age (Nordin et al, 1980). Partial fractures were more frequent, with a prevalence rate of 60% in women 75 years of age and older (Nordin et al, 1980).

In a study of hospital personnel and outpatients, 1.5% of women 50 to 59 years of age and 18% of women 70 to 74 years of age had a partial or complete vertebral fracture (Smith and Rizek, 1966). Studies report a greater prevalence of vertebral fractures in women than in men (Cooper, 1989; Pogrund et al, 1977). To date, there is little information on ethnic variation in the prevalence of vertebral fractures.

Fractures of the distal forearm (Colles' fractures) are the most common fractures among adults until the age of 75 when they are surpassed by hip fractures (Lawrence et al, 1989; Cummings et al, 1985). Table 3 shows incidence data from Rochester, Minnesota (Owen et al, 1982). Women have higher incidence rates than men at all ages over 35. For those women 85 years of age and older the incidence rate is 9 times greater than in men. Cummings et al used the Rochester, Minnesota data to estimate the lifetime risk of a distal forearm fracture among white women (Cummings et al, 1985) and determined that at age 50 a white woman has a 15% risk over the remainder of her life of fracturing her forearm.

Other fractures, those of the humerus and the pelvis for example, are also more common in the older age groups (Lawrence et al, 1989; Cummings et al, 1985; Rose et al, 1982; Melton et al, 1981). A high per-

Table 3: Incidence of First Colles' Fractures Among Adult Residents of Rochester, Minnesota by Age and Gender, 1945-1974

	(Rate per 100,000 persons)	
Age	Males	Females
35-39 years	**43.8**	**106.6**
40-44 years	**70.1**	**116.8**
45-49 years	**78.8**	**194.5**
50-54 years	**118.4**	**355.4**
55-59 years	**92.9**	**494.9**
60-64 years	**74.3**	**639.8**
65-69 years	**113.3**	**537.6**
70-74 years	**-**	**669.8**
75-79 years	**98.9**	**517.8**
80-84 years	**-**	**526.9**
85 years & older	**78.3**	**688.2**
Total	**81.8**	**368.0**

Source: *Owen et al, 1982.*

centage of these fractures, 70% to 80%, occurring in older persons result from minimal trauma and are suggestive of underlying osteoporosis.

Hip fractures

As indicated above, hip fracture is a major consequence of osteoporosis, and because of its frequency among the fastest growing segments of the population, it is a major public health concern. Hip fracture is a significant contributor to morbidity and disability for the older population and is associated with increased mortality in the year after fracture. (Cummings, et al, 1985; Melton et al, 1986; Jacobsen et al, 1990a). Current estimates of hip fractures indicate approximately 250,000 occur per year (Brody, 1985; Brody, 1987). Estimates based on census projections indicate that there will be approximately 340,000 hip fractures per year by the year 2000, and more than 650,000 per year by the year 2050, one-half of which will be in persons 85 years of age and older (Brody, 1985).

The incidence of hip fracture varies directly with age and the incidence rate in a number of studies shows near exponential increases after age 50, with an approximate doubling of the rate for every 5-to 6-year increase in age. (Brody, 1987; Melton et al, 1986; Farmer et al, 1984; Brody et al, 1984).

Rates also show a significant variation by gender, race and ethnic group. A recent study utilizing 1984 through 1987 data from the Health Care Financing Administration and the Department of Veterans Affairs noted that the incidence rates of hip fracture for the 65 and older population vary by gender and racial group. According to Jacobsen et al, 1990, the age-adjusted rate was highest among white women (8.07/1,000), followed by white men (4.28/1,000), black women (3.06/1,000), and black men (2.38/1,000). The incidence rate among white women rose exponentially from 1.63/1,000 at age 65 to 35.4/1,000 at age 95 (Jacobsen et al, 1990a). An exponential rise in rates was also noted for white men, but not for black women or black men (Figure 1).

Figure 1: Hip Fracture Incidence Among the United States Elderly Population Age-, Race-, and Gender-Specific Annual Incidence Rates

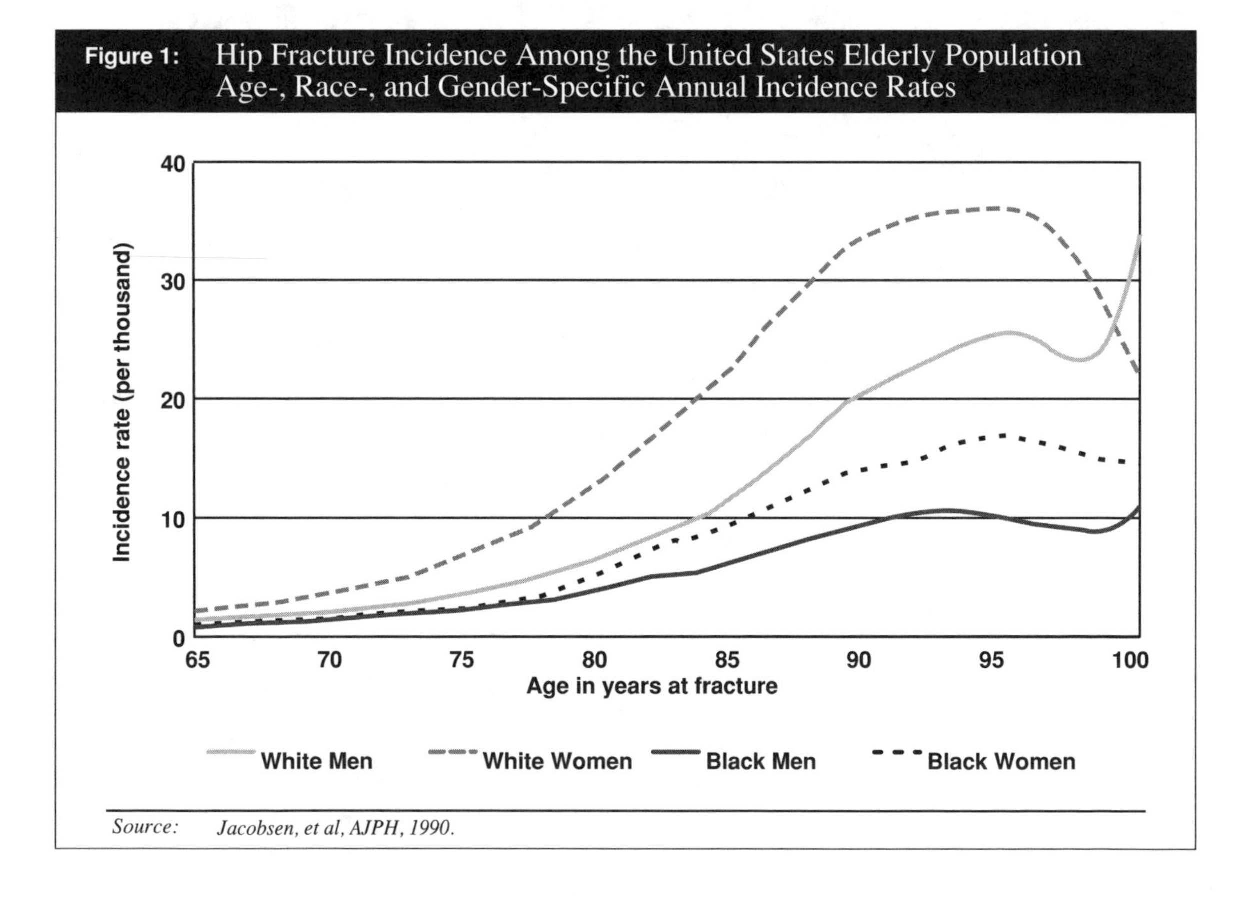

Source: *Jacobsen, et al, AJPH, 1990.*

Ethnic differences in hip fracture incidence have also been noted (Silverman and Madison, 1988; Bauer, 1988; Ross et al, 1991). In particular, Silverman noted that there was lower risk for hip fracture after age 60 in Hispanic (age-adjusted rate of 49.7/100,000), black (57.3/100,000), and Asian-American (85.4/100,000) women than in non-Hispanic white women (140.7/100,000). The exponential rise was again noted, and in this study it was apparent in all four of these ethnic groups. Among men, the highest rates were in non-Hispanic white men but the ethnic difference noted in women was not as evident in men. Bauer, in a study in Bexar County Texas, compared hip fracture incidence rates for non-Hispanic whites, Mexican-Americans, and blacks. He found that the rates were highest among women for the non-Hispanic whites (139/100,000), second highest for Mexican-Americans (67/100,000), and lowest for blacks (55/100,000). Similar patterns were evident for men; however, the rates were approximately one-half the rates for women. Ross et al compared hip fracture incidence among native Japanese, Japanese-American and American Caucasians and noted that the age-specific and cumulative hip fracture rate among persons of Japanese ancestry was approximately half that of Caucasians.

Hip fracture rates have been reported to vary by geographic region (Jacobsen et al, 1990b; Bacon et al, 1989). In the Jacobsen et al study of Health Care Financing Administration data for 1984 through 1987, a North-to-South geographic pattern (higher rates in the South) was noted. Of particular interest was an apparent cluster that extended from the Texas panhandle east to Arkansas, northern Mississippi, Alabama, and Georgia.

Persons who suffer a hip fracture are at increased risk for disability, death and high medical costs (Cummings et al, 1985; Melton et al, 1986; Kellie and Brody, 1990; Sattin et al, 1990).

Impact of hip fracture

Of those who survive a hip fracture, as many as half spend time in a long-term care institution (Cummings et al, 1985; Kellie and Brody, 1990; Sattin et al, 1990). Of those who were independent and living at home when the fracture occurred, about 20% remained in a long-term care institution for at least a year after fracture (Cummings et al, 1985). Not surprisingly, there was excess risk of institutionalization among those with pre-existing impairments and multiple chronic conditions. Although 30% of those who were independent and living at home at time of fracture do return home, they must depend on other people or assistive devices for mobility.

The mortality rate in the first year after fracture is about 15% higher than in persons of comparable age and gender who have not had a hip fracture (Cummings et al, 1985). Most of these deaths occur in the first four months after fracture.

References

Bacon WE, Smith GS, Baker SP. Geographic Variation in the Occurrence of Hip Fractures Among the Elderly White US Population. *Am J Public Health*, 79:1556-1558, 1989.

Bauer R. Ethnic Differences in Hip Fracture. A Reduced Incidence in Mexican Americans. *Am J Epidemiol*, 127:145-149, 1988.

Brody JA. Prospects for an Aging Population. *Nature*, Vol. 315: No. 6019, pp. 463-466, 1985.

Brody JA. Aging in the 20th and 21st Centuries. In: *Proceedings of the National Center for Health Statistics Conference 'Data for an Aging Population Issues in Health Research and Public Policy, Now and Into the 21st Century'*, DHHS Pub. No. (PHS) 88-1214, 1987.

Brody JA, Farmer ME, Whit, LR. Absence of Menopausal Effect on Hip Fracture Occurrence in White Females. *Am J Publ Health*, 74:1397-1398, 1984.

Cooper C. Osteoporosis–an Epidemiologic Perspective: A Review. *J Royal Society of Med*, 82:753-757, 1989.

Cummings SR, Kelsey JL, Nevitt MC, O'Dowd KJ. Epidemiology of Osteoporosis and Osteoporotic Fractures. *Epi Reviews*, Vol. 7, 1985.

Farmer ME, White LR, Brody JA, Bailey KR. Race and Sex Differences in Hip Fracture Incidence. *Am J Publ Health*, 74:1374-1380, 1984.

Graham, D. Detailed Diagnoses and Procedures for Patients Discharged from Short-stay Hospitals United States, 1986. *Vital and Health Statistics,* Series 13, No. 95, DHHS, 1988.

Iskrant AP, Smith R, Jr. Osteoporosis in Women 45 Years and Older Related to Subsequent Fractures. *Public Health Rep*, 84:33-38, 1969.

Jacobsen, SJ, Goldberg, J, Miles, TP, Brody, JA, Stiers, W, Rimm, AA. Hip Fracture Incidence Among the Old and Very Old: A Population-based Study of 745,435 cases. *Am J Public Health*, 80:871-873, 1990a.

Jacobsen SJ, Goldberg J, Miles TP, Brody W, Rimm AA. Regional Variation in the Incidence of Hip Fracture, US White Women Aged 65 Years and Older. *JAMA*, Vol. 264, No. 4:500-502, 1990b.

Jacobsen SJ, Goldberg J, Miles TP, Brody JA, Stiers W, Rimm AA. Seasonal Variation in the Incidence of Hip Fracture Among White Persons Aged 65 Years and Older in the United States.

Jensen GR, Christiansen C, Boesen J, et al. Epidemiology of Postmenopausal Spinal and Long Bone Fractures—a Unifying Approach to Postmenopausal Osteoporosis. *Clin Orthop*,166:75-81, 1982.

Kellie, SE, Brody, JA. Sex-specific and Race-specific Hip Fracture Rates. *Am J Public Health*, 80:326-328, 1990.

Lawrence RC, Hochberg MC, Kelsey JL, McDuffie FC, Medsger TA, Felts WR, Shulman LE. Estimates of the Prevalence of Selected Arthritic and Musculoskeletal Diseases in the United States. J Rheumatol,16(4):427-441, 1989.

Lewinnnek G, Kelsey JL, White AA, et al. The Significance and Comparative Analysis of the Epidemiology of Hip Fractures. *Clin Orthop*, 152:35-43, 1980.

Mazess, RB. On Aging Bone Loss. Clin Orthop, 165:239-252, 1982.

Melton LJ, Wahner HW, Michelson LS, O'Fallon WM, Riggs BL. Osteoporosis and the Risk of Hip Fracture. *Am J Epidemiol*, 124:254-261, 1986.

Melton LJ III, Jampson JM, Morrey BF, et al. Epidemiologic Features of Pelvic Fractures. *Clin Orthop*, 155:43-47, 1981.

Nordin BEC, Peacock M, Aaron J, Crilly RG, Heyburn PJ, Horsman A, Marshal, D. Osteoporosis and Osteomalacia. *Clin Endocrinol Metab*, 9:177-204, 1980.

Owen RA, Melton LJ III, Johnson KA, et al. Incidence of Colles' Fracture in a North American Community. *Am J Public Health*, 72:605-607, 1982.

Pogrund H, Makin M, Robin G, Menczel J, Steinberg R. Osteoporosis in Patients with Fractured Femoral Neck in Jerusalem. *Clin Orthop*, 124:165-172, 1977.

Riggs BL, Wahner HW, Dunn WL, Mazess RB, Offord KP, Melton LJ. Differential Changes in Bone Mineral Density of the Appendicular and Axial Skeleton with Aging. *J Clin Invest,* 67:328-335, 1981.

Rose SH, Melton LJ III, Morrey BF, et al. Epidemiologic Features of Humeral Fractures. *Clin Orthop*, 168:24-30, 1982.

Sattin RW, Lambert-Huber DA, DeVito CA, Rodriquez JG, Ros A, Bacchelli S, Stevens JA, Waxweiler RJ. The Incidence of Fall Injury Events Among the Elderly in a Defined Population. *Am J Epidemiology*, 131, 1990.

Silverman SL, Madison RE. Decreased Incidence of Hip Fracture in Hispanics, Asians, and Blacks: California Hospital Discharge Data. *Am J Public Health*, 78:1482-1483, 1988.

Smith RW, Rizek J. Epidemiologic Studies of Osteoporosis in Women of Puerto Rico and Southeastern Michigan with Special Reference to Age, Race, National Origin and Other Related or Associated Findings. *Clin Orthop*, 45:31-48, 1966.

chapter 4

Neoplasms of Bone and Connective Tissue

Bone and connective tissue neoplasms are rare when compared with other cancers and with other musculoskeletal conditions. Although data indicate that there has been improvement in the prognosis of some histologic categories of these neoplasms (Young et al, 1986; Goorin et al, 1985), little progress has been made toward understanding their etiology (Fraumeni and Boice, 1980).

Fifteen years of data from the Surveillance, Epidemiology and End Results (SEER) program of the National Cancer Institute have been used to examine the descriptive epidemiology of these malignancies. Data are provided on 2,727 bone cancers, 6,883 connective tissue tumors, and 12,945 multiple myelomas. The SEER program is the most comprehensive source of neoplasm data and is based on data representing approximately 10% of the United States population.

Incidence

Figures 1 through 3 show the age-adjusted incidence rates for bone, connective tissue, and multiple myeloma neoplasms from 1973 to 1987. Incidence rates appear to have been relatively stable over this period,

Figure 1: Bone Cancer, Age Adjusted Incidence Rates* by Year

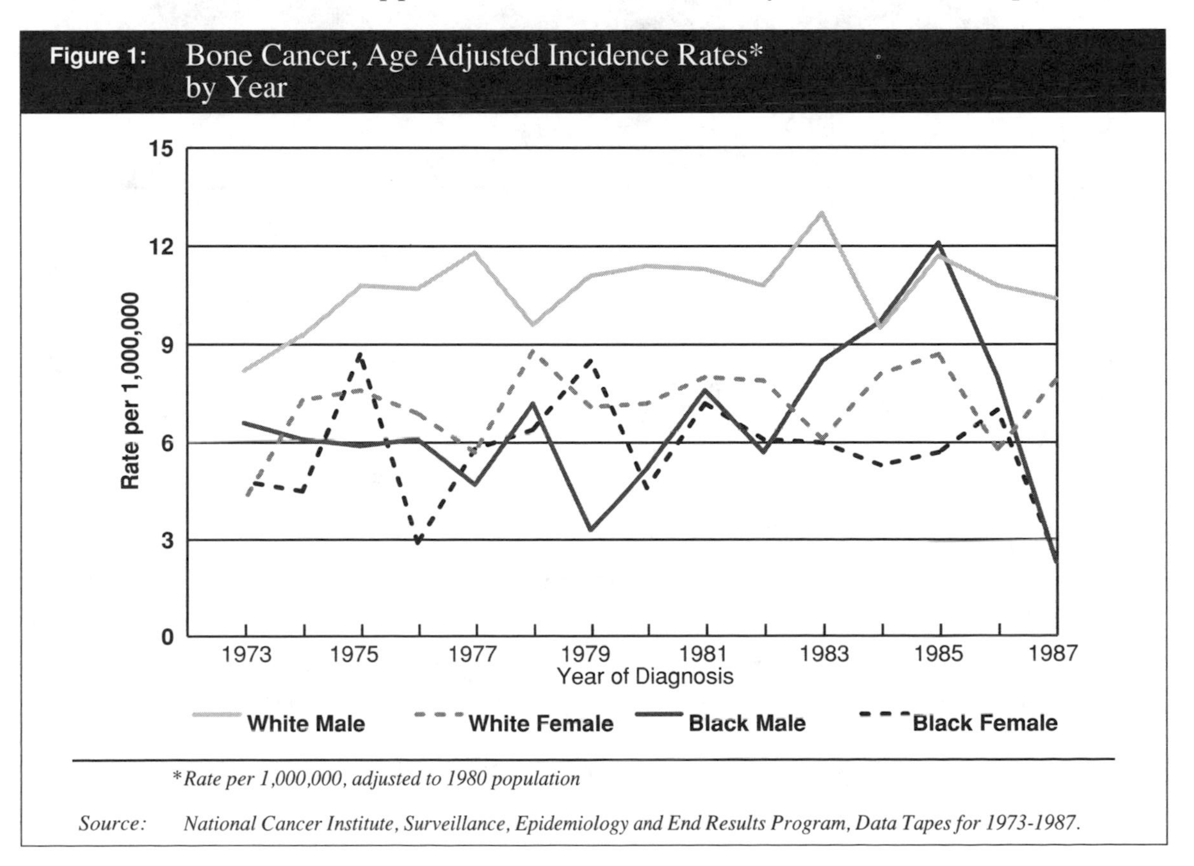

*Rate per 1,000,000, adjusted to 1980 population

Source: National Cancer Institute, Surveillance, Epidemiology and End Results Program, Data Tapes for 1973-1987.

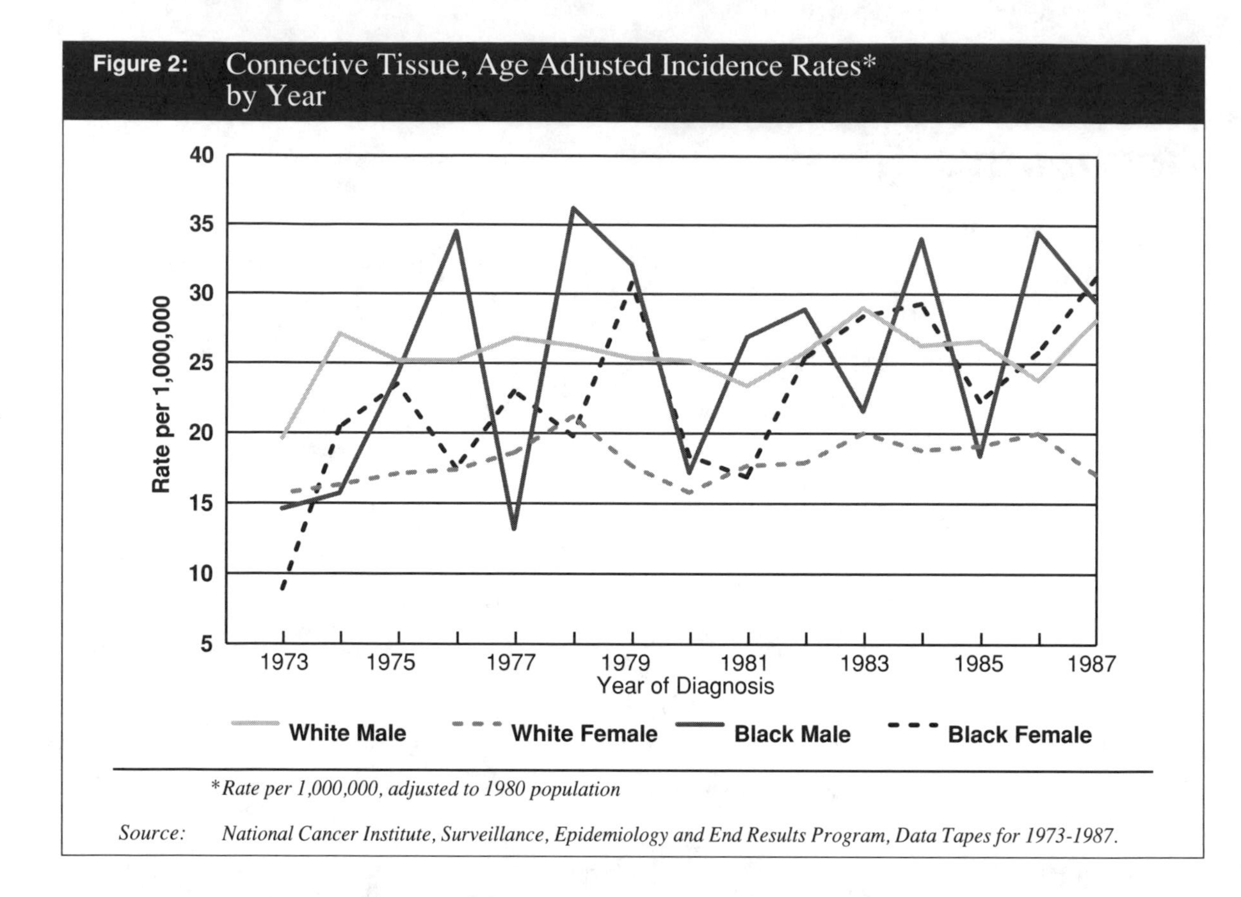

Figure 2: Connective Tissue, Age Adjusted Incidence Rates* by Year

*Rate per 1,000,000, adjusted to 1980 population

Source: *National Cancer Institute, Surveillance, Epidemiology and End Results Program, Data Tapes for 1973-1987.*

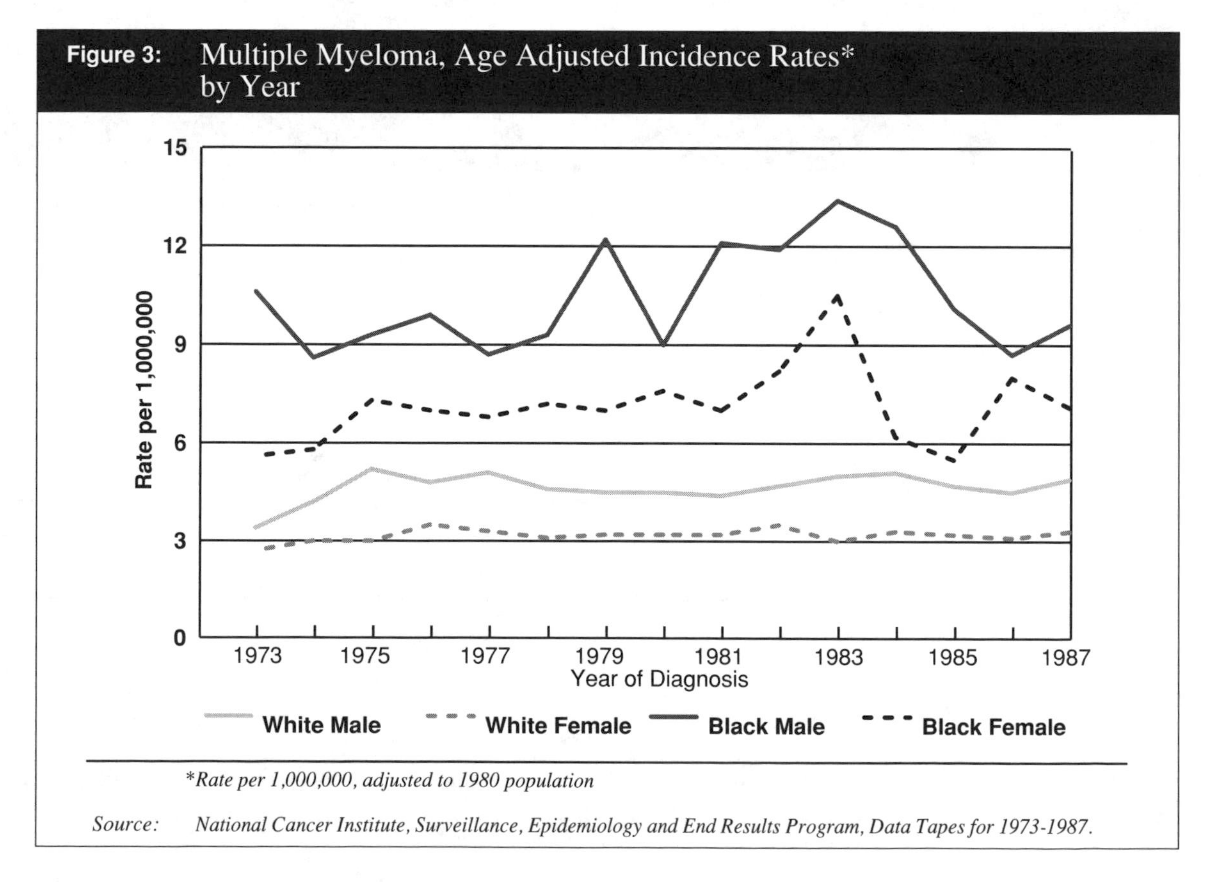

Figure 3: Multiple Myeloma, Age Adjusted Incidence Rates* by Year

*Rate per 1,000,000, adjusted to 1980 population

Source: *National Cancer Institute, Surveillance, Epidemiology and End Results Program, Data Tapes for 1973-1987.*

with the exception of connective tissue neoplasms which show a slight upward trend. The greater year-to-year variability among incidence rates for blacks reflects the smaller population on which these rates are based.

Current data indicate that multiple myeloma is the most frequently occurring of these malignancies, followed by connective tissue and bone cancers (Table 1). Among bone cancers, osteosarcoma, chondrosarcoma,

Table 1: Age-adjusted Incidence Rates for 1987 for Selected Neoplasms by Race and Gender

(Rate per 1,000,000 persons)

	White Male	White Female	Black Male	Black Female
Bone	10.40	7.92	2.32	2.36
Osteosarcoma	3.50	3.27	1.57	0.76
Chondrosarcoma	2.73	1.96	0.00	0.93
Ewing's Sarcoma	2.13	1.57	0.75	0.00
Connective Tissue	28.13	17.07	29.46	31.17
Multiple Myeloma	49.03	33.33	96.39	70.60

Source: *National Cancer Institute, Surveillance, Epidemiology and End Results Program, Data Tapes for 1987.*

and Ewing's sarcoma are the most frequent and account for approximately 83% of all bone cancers.

Figures 4 and 5 show the age-specific incidence rates of all bone and connective tissue malignancies by race and gender. Rates for bone cancers in males tend to be higher than for females. Rates among whites exhibit a bimodal distribution that peaks at ages 15 to 20 and again after 70 years of age (Figure 4). The young adult peak in incidence is also apparent in blacks, with rates which are similar to whites. The older peak in incidence is less evident in blacks and the incidence rates in older blacks are lower than in whites. This pattern is consistent with previous reports (Fraumeni and Boice, 1980). In contrast to bone cancers, connective tissue neoplasms occur with greater overall frequency and incidence rates steadily increase with age. Rates in males tend to be higher than in females and, with the exception of the very old, rates are similar between blacks and whites (Figure 5).

Figure 4: Bone Cancer, Age Specific Incidence Rates*

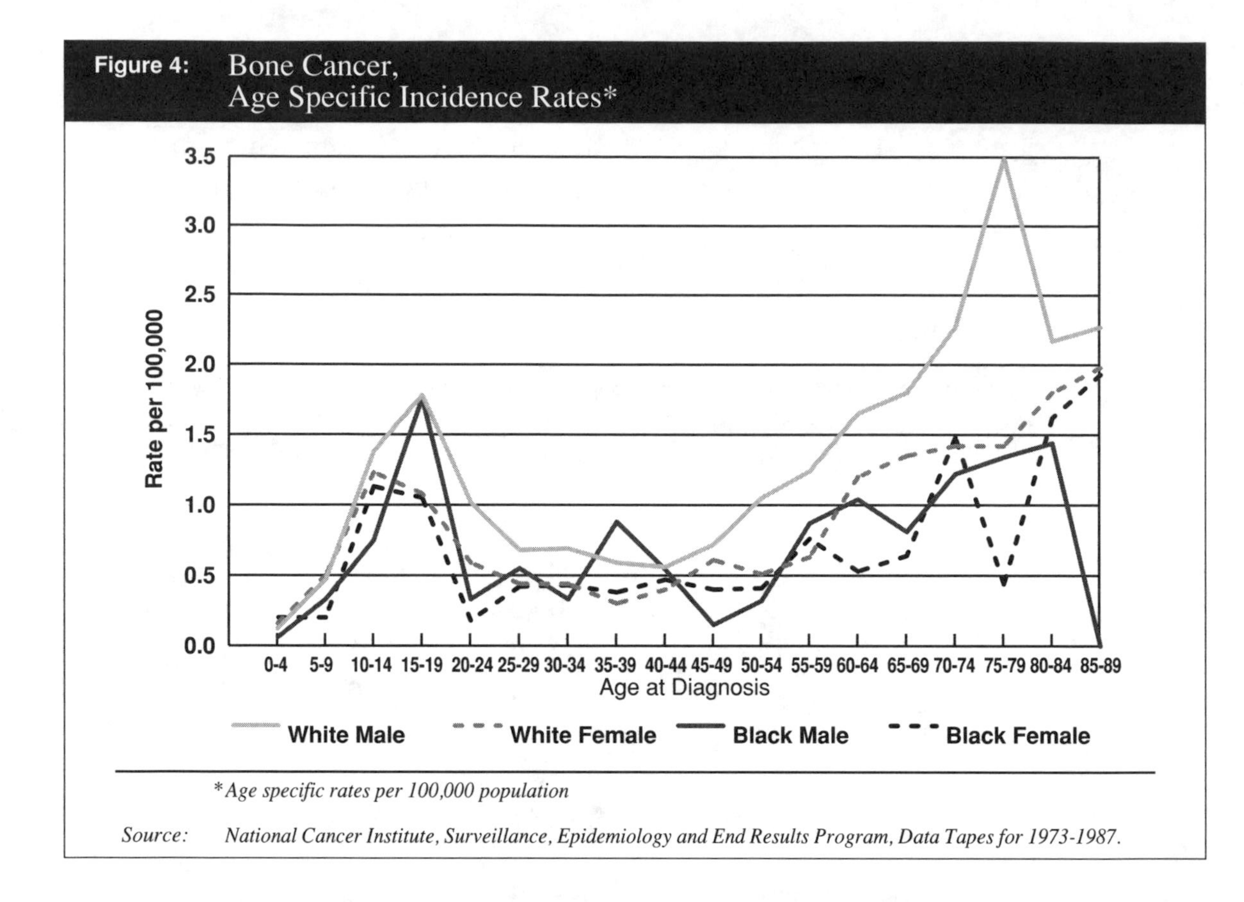

**Age specific rates per 100,000 population*

Source: *National Cancer Institute, Surveillance, Epidemiology and End Results Program, Data Tapes for 1973-1987.*

Figure 5: Connective Tissue, Neoplasms, Age Specific Incidence Rates*

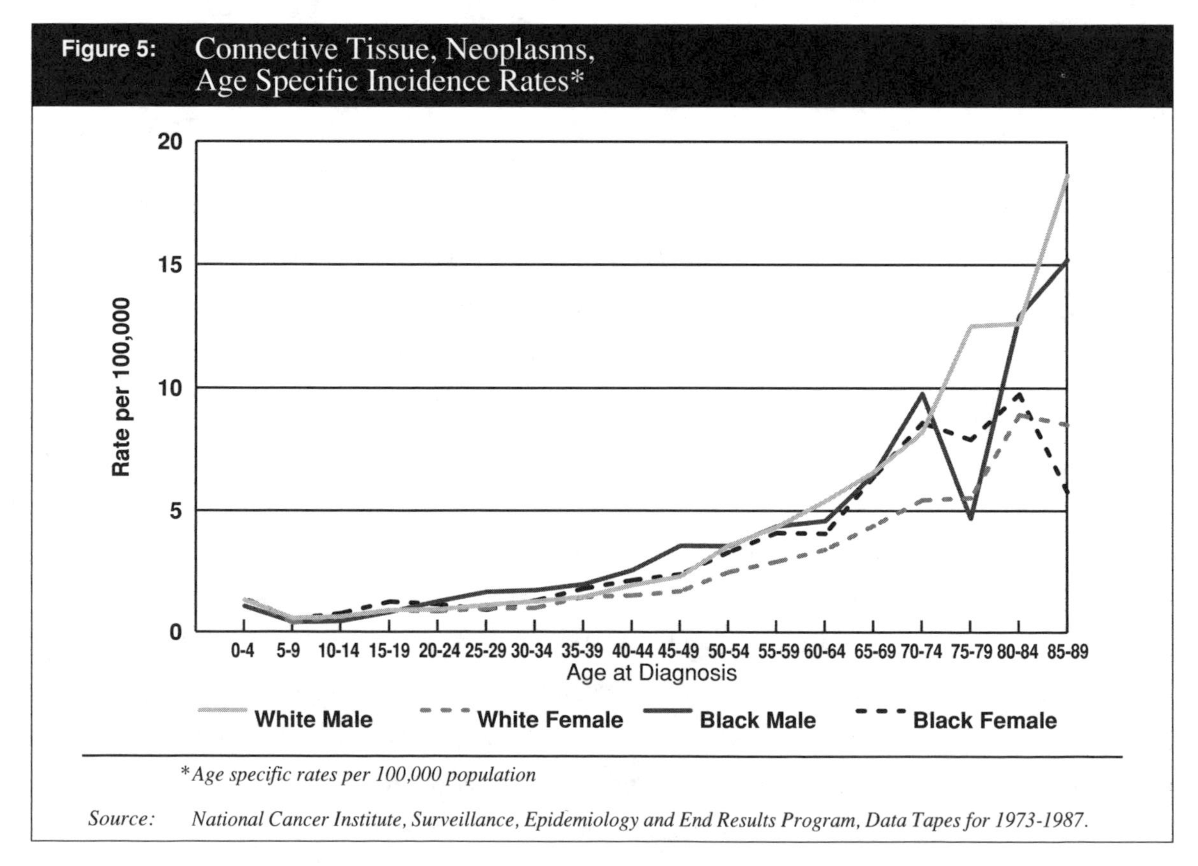

**Age specific rates per 100,000 population*

Source: *National Cancer Institute, Surveillance, Epidemiology and End Results Program, Data Tapes for 1973-1987.*

The incidence rate of multiple myeloma increases markedly for both blacks and whites with age, with rates for blacks approximately twice as high as rates for whites in the older age groups (Figure 6). Rates are higher among males for both whites and blacks.

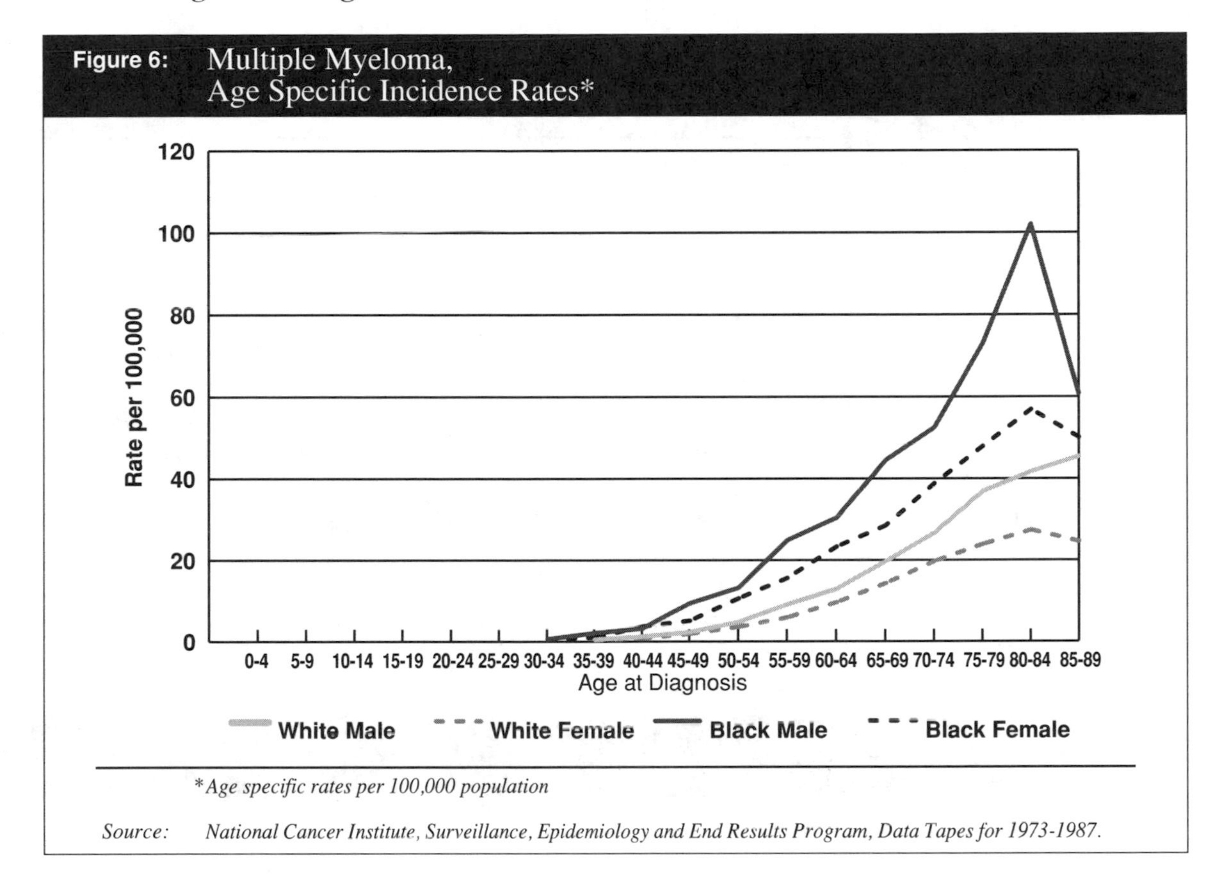

Figure 6: Multiple Myeloma, Age Specific Incidence Rates*

**Age specific rates per 100,000 population*

Source: *National Cancer Institute, Surveillance, Epidemiology and End Results Program, Data Tapes for 1973-1987.*

The pattern of incidence for subtypes of bone cancer varies markedly. In whites, osteosarcoma has a bimodal age distribution; blacks experience the peak at younger ages only (Figure 7). Chondrosarcomas tend to increase with age (Figure 8), while Ewing's sarcoma occurs primarily in whites under the age of 30 (Figure 9). These age and racial patterns are similar to those reported in the state of New York (Polednek, 1985).

Survival experience

Black and white males have a similar five-year survival experience with respect to bone, connective tissue and multiple myeloma (Table 3). The similarity in survival rates by race for osteosarcoma is consistent with previous reports (Huvos et al, 1983). Women have a better five-year survival experience for osteosarcoma, a pattern that has previously been

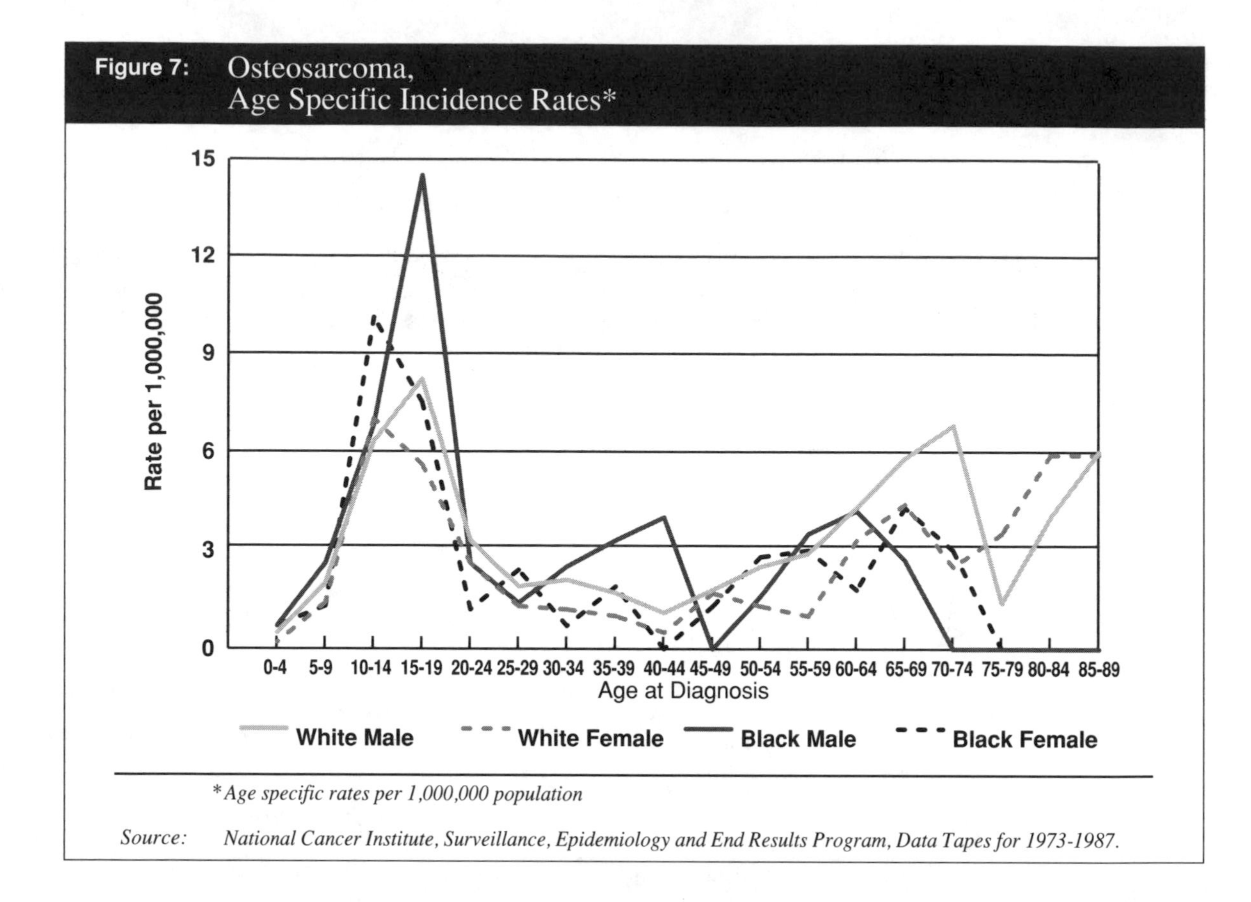

Figure 7: Osteosarcoma, Age Specific Incidence Rates*

**Age specific rates per 1,000,000 population*

Source: *National Cancer Institute, Surveillance, Epidemiology and End Results Program, Data Tapes for 1973-1987.*

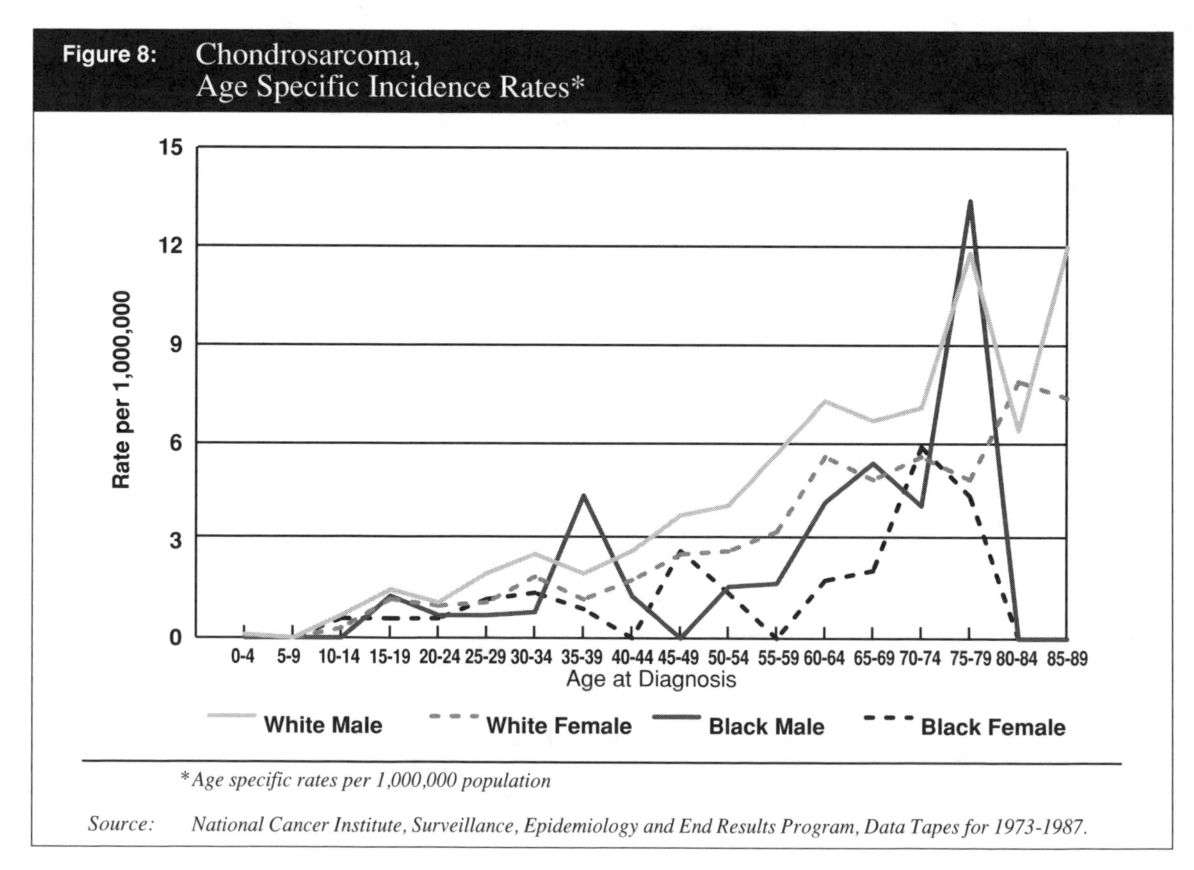

Figure 8: Chondrosarcoma, Age Specific Incidence Rates*

**Age specific rates per 1,000,000 population*

Source: *National Cancer Institute, Surveillance, Epidemiology and End Results Program, Data Tapes for 1973-1987.*

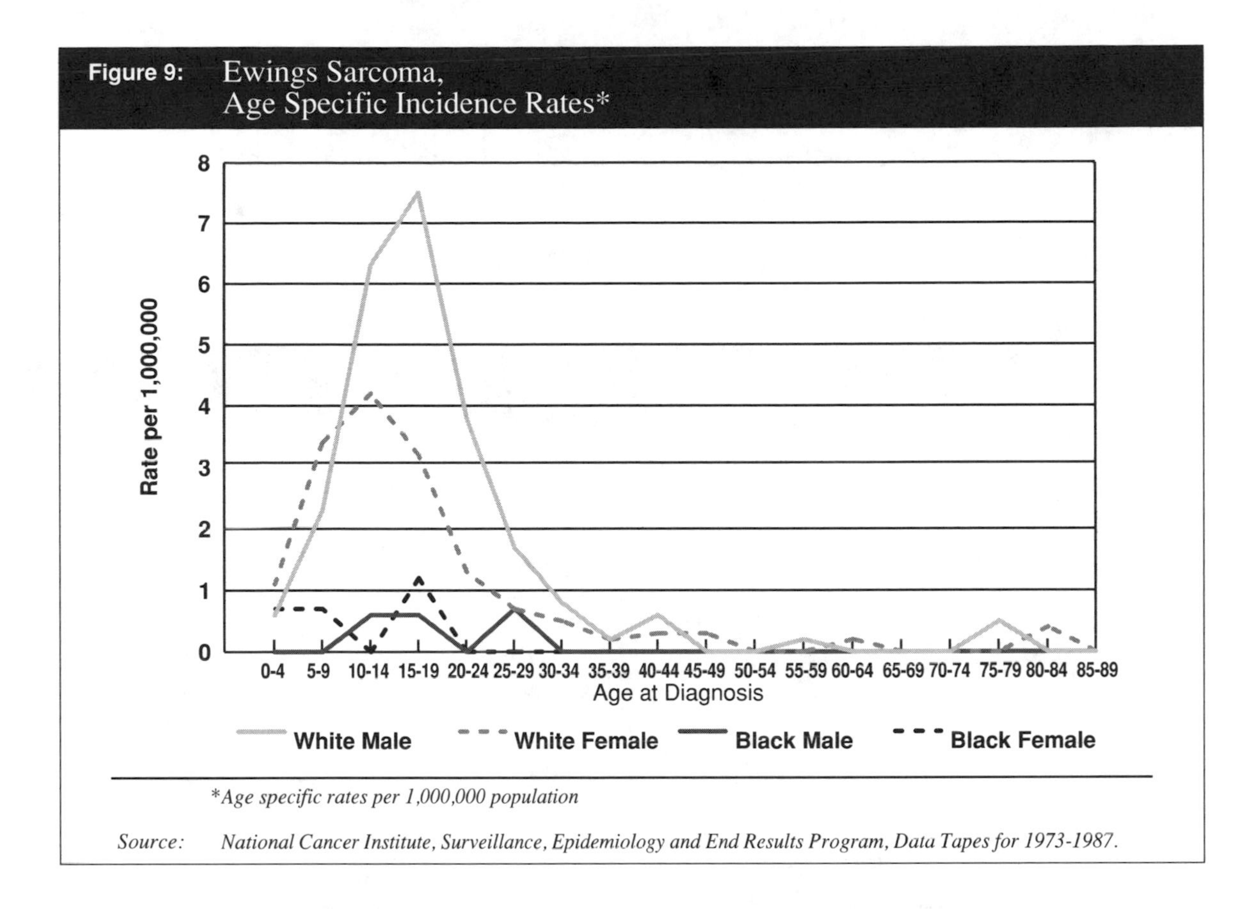

*Age specific rates per 1,000,000 population

Source: National Cancer Institute, Surveillance, Epidemiology and End Results Program, Data Tapes for 1973-1987.

Table 3: Crude Five-year Survival Rates for Cases Diagnosed Between 1973 and 1983

(Percent of cases)

	White Male	White Female	Black Male	Black Female
Bone	38	42	36	46
Osteosarcoma	28	37	29	43
Chondrosarcoma	59	53	43	83
Ewing's Sarcoma	28	42	100	33
Connective Tissue	43	41	43	45
Multiple Myeloma	15	18	17	22

Source: National Cancer Institute, Surveillance, Epidemiology and End Results Program, Data Tapes for 1973-1987.

observed in young adult cases (Homa, et al, 1991). White women with Ewing's sarcoma have a better survival experience than white men. The survival rates in blacks with Ewing's sarcoma are likely to demonstrate an erratic pattern because of the small number of cases. Median survival in months is indicated in Table 4.

Table 4: Median Survival for Cases Diagnosed Between 1973 and 1983

	(Survival in months)			
	White Male	White Female	Black Male	Black Female
Bone	**38**	**50**	**37**	**49**
Osteosarcoma	**22**	**30**	**27**	**49**
Chondrosarcoma	**71**	**53**	**43**	**101**
Ewing's Sarcoma	**29**	**50**	**113**	**18**
Connective Tissue	**51**	**46**	**54**	**54**
Multiple Myeloma	**20**	**21**	**21**	**25**

Source: National Cancer Institute, Surveillance, Epidemiology and End Results Program, Data Tapes for 1973-1987.

Etiology

Little is known about the etiology of bone and connective tissue tumors, but the varying incidence patterns suggest that they arise from different etiologic processes. Ionizing radiation has been clearly established as an agent that produces bone tumors, primarily osteogenic sarcoma, chondrosarcoma and fibrosarcoma. Tumors seem to arise from high doses of radiation, specifically bone-seeking radium isotopes (RA2267, RA224), external radiation and internally deposited radioisotopes. The magnitude of the effect appears to depend on the type and level of exposure, but in general, radiation exposure accounts for a small proportion of the cases (Fraumeni and Boice, 1980).

Other factors, such as exposure to chemicals and viruses, have been implicated in the development of bone tumors in animals, but there is little evidence to support this relationship in humans. Specifically, beryllium, vinyl chloride, procarbazine (an antitumor agent), aflatoxin and trace metals have been implicated. A recent study in New York concluded that an earlier association between fluoride in drinking water and the occurrence of bone cancer could not be supported (Mahoney et al, 1991).

Paget's disease predisposes to osteogenic sarcoma, fibrosarcoma, chondrosarcoma and giant cell tumors, and men with Paget's disease are twice as likely as women with Paget's disease to develop cancer. A familial tendency for osteogenic tumor, more than for chondrosarcoma or Ewing's sarcoma, has been suggested by observations of multiple cases in families.

Persons treated for bone cancer are at increased risk of developing a second bone malignancy. Both radiotherapy and alkylating agents may serve to increase this potential.

The high incidence of osteosarcomas during adolescence has been correlated with skeletal growth patterns. The peak incidence of osteogenic sarcoma during adolescence has led to the hypothesis that bone tumors may be a pathologic response to the stimulus of anabolic steroid hormones during puberty. A study has demonstrated that estrogen can act directly on osteoblasts by a receptor-mediated mechanism. In doing so, the extracellular matrix and other proteins involved in the maintenance of skeletal mineralization and remodeling may be modulated (Komm et al, 1988). Although few endocrinologic studies have been completed, one did show abnormal glucose, insulin and growth hormone responses to glucose stimulation, which raises the possibility that growth factors may influence the development of osteosarcoma (Fraumeni and Boice, 1980).

The rarity of bone and connective tissue tumors has impeded progress in developing an understanding of the risk factors for these malignancies. An increasing number of population-based state cancer registries should result in a greater number of collaborative clinical and epidemiologic studies.

References

Fraumeni JF and Boice JD. Bone. In Schottenfeld, ed, *Cancer Epidemiology* 1980.

Goorin AM, Abelson HT and Frei E. Osteosarcoma: Fifteen Years Later. *NEJM* 313:26;1637-1643, 1985.

Homa DM, Sowers MR and Schwartz AG. Incidence and Survival Rates of Children and Young Adults with Osteogenic Sarcoma. *Cancer* 67:2219-2223, 1991.

Huvos AG, Butler A and Bretsky S. Osteogenic Sarcoma in the American Black. *Cancer* 52:1959-1965, 1983.

Komm BS, Terpening CM, Benz DJ et al. Estrogen Binding, Receptor mRNA, and Biologic Response in Osteoblast-like Osteosarcoma Cells. *Science* 241:81-84, 1988.

Mahoney MC, Nasca PC, Burnett WS and Melius JM. Bone Cancer Incidence Rates in New York State: Time Trends and Fluoridated Drinking Water. *Am J Public Health* 81:475-479, 1991.

National Cancer Institute. Surveillance, Epidemiology and End Results Program, 1973-1987 (Data Tapes).

Polednek AP. Primary Bone Cancer Incidence in Black and White Residents of New York State. *Cancer* 55:2883-2888, 1985.

Taylor WF, Ivins JC, Unni KK et al. Prognostic Variables in Osteosarcoma: A Multi-Institutional Study. *Journal of the National Cancer Institute* 81:21-30, 1989.

Young JL, Gloeckler Ries L, Silverberg E. et al. Cancer Incidence, Survival, and Mortality for Children Younger than Age 15 Years. *Cancer* 58:598-602, 1986.

chapter 5

Congenital Malformations of the Musculoskeletal System

Congenital malformations of the musculoskeletal system include a variety of anomalies and range from minor malformations such as extra fingers or toes to serious disabling conditions such as spina bifida. The primary source of data relating to congenital malformations is the Centers for Disease Control's (CDC) Birth Defects Monitoring Program (BDMP). The BDMP is a national program that monitors congenital malformations by using hospital discharge data concerning newborns. The program's database contains records on more than 15 million births occurring since 1970 and collects data each year on more than 15% of United States births.

Prevalence (at birth)

Prevalence at birth rates for selected musculoskeletal anomalies for the years 1979-80 and 1986-87 are indicated in Table 1. In 1986-87, clubfoot without central nervous system (CNS) defects occurred at a rate of 25.11/10,000 births. Spina bifida without anencephalus occurred at a rate of 4.45/10,000 total births; reduction deformities of the upper and lower limbs occurred at rates of 1.58 and 0.83/10,000 births respectively. Congenital arthrogryposis (amyoplasia congenita) occurred at a rate of 1.93/10,000 births.

Table 1: Selected Congenital Malformations: United States 1979-80 and 1986-87

	(Rate per 10,000 Total Births)	
	1979-80	1986-87
Spina bifida, without anencephalus	**5.11**	**4.45**
Clubfoot, without CNS defects	**25.62**	**25.11**
Reduction deformity, upper limbs	**1.53**	**1.58**
Reduction deformity, lower limbs	**0.78**	**0.83**
Congenital arthrogryposis	**1.33**	**1.93**

Source: MMWR, Vol. 39, SS-4, 1990.

Based on these data, estimates in Table 2 indicate the number of infants born with these conditions in the United States in 1986 and 1987. The two most frequently occurring were clubfoot and spina bifida without anencephalus occurring in approximately 9,600 and 1,700 infants respectively each year.

Table 2: Estimated Number of Infants Born per Year in the United States with Selected Musculoskeletal Malformations: 1986-87*

	1986-87
Spina bifida, without anencephalus	**1,700**
Clubfoot, without CNS defects	**9,600**
Reduction deformity, upper limbs	**600**
Reduction deformity, lower limbs	**300**
Congenital arthrogryposis	**800**

**Estimates based on MMWR, Vol. 39, SS-4, 1990.*

Comparing rates in the different time periods (Table 1), indicates relatively stable rates for clubfoot and reduction deformities. Congenital arthrogryposis rates, however, increased 45% from 1.33/10,000 in 1979-80 to 1.93/10,000 in 1986-87. Spina bifida showed a 12.9% decline from 5.11 to 4.45/10,000. This decline appears to be part of a continuing trend with prevalence at birth rates of spina bifida without anencephalus declining from levels of approximately 7.9/10,000 in the early 1970's. Maternal lifestyle changes, including better nutrition, were suggested as possible factors in the decline in rates for spina bifida (MMWR, Vol. 39, SS-4). Terminations of pregnancy may be another factor.

Spina bifida is the most common of the severe congenital anomalies and is thought to result from the failure of the posterior part of the vertebral column to fuse during early fetal development. Spina bifida can range in severity from a relatively minor condition without symptoms that may be diagnosed incidentally during a radiographic examination of the spine (spina bifida occulta), to a severe defect involving the spinal canal in which the contents of the canal are exposed on the body surface (spina bifida with meningomyelocele). Although rates for spina bifida have declined, it is still a significant health concern, with the BDMP estimating that between 1980 and 1987, more than 13,600 infants were born with this condition (MMWR, Vol. 38, No. 15).

Congenital malformations show a considerable variation by ethnic group. Using BDMP data for the period from 1981 through 1986, the CDC determined rates for the 18 most frequent major congenital malformations occurring within each of five ethnic and racial groups (MMWR 1988, Vol. 37, SS-3: 17-24). Fifteen malformations were found to be common to all groups. Rates for three congenital musculoskeletal malformations identified as among the 15 most frequent are listed in Table 3.

Table 3: Rates of Selected Congenital Malformations by Race and Ethnicity: United States, 1981-86

(Rate per 10,000 Total Births)

	Blacks	Hispanics	American Indians	Asians	Whites
Spina bifida without anencephalus	**3.3**	**5.9**	**4.1**	**1.8**	**5.1**
Clubfoot without CNS defects	**19.9**	**19.1**	**15.5**	**14.4**	**27.5**
Hip dislocation without CNS defects	**13.8**	**24.0**	**31.4**	**25.0**	**32.3**

Source: *MMWR, Vol. 37, SS-3, 1988.*

Aggregating the BDMP data over the 1981-86 period, prevalence rates for spina bifida are more than three times greater in Hispanics (5.9/1,000 births) than in Asian-Americans (1.8/10,000 births). Other musculoskeletal malformations, clubfoot without CNS defects and hip dislocation without CNS defects, also show varying rates by ethnic group, although the percent variation is smaller. For all three conditions, however, prevalence rates for whites are above average and tend to be lowest for Asian-Americans and Blacks.

Limited regional analysis is available primarily because of the comparatively small number of congenital malformations compared with other conditions. A study by Greenberg and associates found that spina bifida rates in the United States were highest in Appalachia and were approximately 10/10,000 live births (Greenberg et al, 1983).

Impact

The impact of congenital malformations of the musculoskeletal system is considerable. Many malformations require extensive medical treatment, sometimes beginning shortly after birth. With attendant complications and limitations, they may continue to affect individuals throughout their lives. These anomalies are also a significant source of infant mortality.

In recent years there has been an increase in survival rates for infants born with spina bifida (MMWR, Vol. 38, No. 15), which may have resulted from advances in surgical techniques. The impact of this malformation among those surviving can be severe. Damage to the spinal cord associated with spina bifida can produce severe disabilities that require extensive and continuing medical and surgical care. Many children are confined to wheelchairs; others may walk with the use of braces. These children may require specialized school facilities and accommodations.

Annual costs for medical and surgical care for all persons in the United States with spina bifida alone have been estimated to exceed $200 million (MMWR, Vol. 38, No. 15). For a person with "typical spina bifida," lifetime costs, which include direct costs for medical and surgical care, long-term care, disability and education, and indirect costs, such as lost productivity, are estimated to total $250,000 in constant (1985) dollars (MMWR, Vol. 38, No. 15).

For other congenital defects, such as clubfoot and congenital dislocation of the hip, the impact upon affected infants and their families is also considerable, although the prognosis for children affected with these conditions is usually more favorable. A favorable prognosis generally depends on early diagnosis and treatment, in the absence of which the deformity may persist into adult life and increase the likelihood of permanent disability. The disability caused by these defects has been reduced because of improvements in treatment, although other conditions, such as osteoarthritis, frequently develop in the affected joints later in life, even with early diagnosis and treatment.

Congenital malformations, including those of the musculoskeletal system, when especially severe, can also be a significant source of infant deaths. As indicated in Table 4, musculoskeletal birth defects account for 8.4% of infant deaths when birth defects were listed as the underlying cause. Including spina bifida as a musculoskeletal condition would make congenital malformations of the musculoskeletal system the third most frequent category of birth defects resulting in infant death.

Table 4: Birth Defects as Underlying Cause of Infant Death, by Organ System: United States, 1986

	Number	Percent
Cardiovascular system	**3,198**	**41.7**
Central nervous system	**1,271**	**16.6**
Chromosomal anomalies	**721**	**9.4**
Musculoskeletal system	**642**	**8.4**
Respiratory system	**498**	**6.5**
Genitourinary system	**458**	**6.0**
Digestive system	**146**	**1.9**
Other	**744**	**9.7**
Total	**7,678**	**100.0**

Source: *MMWR, Vol. 39, SS-3.*

Infant mortality rates by race for spina bifida and congenital anomalies of the musculoskeletal system are indicated in Table 5.

As was the case for the prevalence of selected musculoskeletal malformations, which was examined earlier in this section, infant mortality is higher among whites than other racial groups for spina bifida as well as for musculoskeletal system anomalies in general.

Table 5: Infant Mortality Rates for Selected Congenital Anomaly Categories, Annual Average 1985-1988, by Race

(Rate per 100,000 live births)

			All Other	
	All Races	White	Total	Black
Infant deaths, Total	**1,025.5**	**884.7**	**1,547.7**	**1,791.9**
Congenital anomalies*	**216.1**	**216.9**	**212.9**	**225.1**
Spina bifida**	**2.1**	**2.3**	**1.6**	**1.9**
Musculoskeletal system***	**15.7**	**16.4**	**13.1**	**14.6**

**ICD-9 Codes 740-759*
***ICD-9 Code 741*
****ICD-9 Codes 754-756*

Source: National Center for Health Statistics: Vital Statistics of The United States, Volume II-Mortality, 1985-1988.

References

Centers for Disease Control. Leading Major Congenital Malformations Among Minority Groups in the United States, 1981-1986. *MMWR*, Vol. 37, No. SS-3, 1988.

Centers for Disease Control. Temporal Trends in the Prevalence of Congenital Malformations at Birth Based on the Birth Defects Monitoring Program, United States, 1979-1987. *MMWR*, Vol. 39, No. SS-4, 1990.

Centers for Disease Control. Contributions of Birth Defects to Infant Mortality Among Racial/Ethnic Minority Groups, United States, 1983. *MMWR*, Vol. 37, No. SS-3, 1988.

Centers for Disease Control. Contributions of Birth Defects to Infant Mortality - United States, 1986. *MMWR*, Vol. 38, 633-635, 1989.

Centers for Disease Control. Estimated Years of Potential Life Lost Before Age 65 and Cause-Specific Mortality, by Cause of Death - United States, 1985. *MMWR*, Vol. 36, 313, 1987

Greenberg, F, James, LM, Oakley, GP Jr. Estimates of Birth Prevalence Rates of Spina Bifida in the United States from Computer-generated Maps. *Am J Obstet Gynecol*, 145:570-573, 1983.

National Center for Health Statistics. Vital Statistics of the United States. Editions for years 1985-88.

chapter 6

Health Care Utilization Selected Conditions

Introduction

Musculoskeletal conditions necessitate extensive use of health care resources because of their prevalence in the population as well as the pain and disability caused by many of these conditions. The average annual number of discharges, inpatient days and average lengths of stay for selected musculoskeletal conditions are indicated in Table 1.

In the period from 1985 through 1988, conditions related to back pain (including back injury) resulted in an average of 886,000 hospitalizations and 6 million inpatient days per year. Conditions associated with back pain accounted for a greater number of hospitalizations than any other individual musculoskeletal condition.

Hospitalizations

Disk disorders accounted for almost half (46.7%) of hospitalizations for conditions related to back pain. Other back disorders such as inflammatory spondylopathies, spondylosis and allied disorders, and other non-disk disorders accounted for 289,000 hospitalizations (32.6%). The average length of stay for conditions related to back pain was 6.7 days and was higher for women in the aggregate and in all four back pain categories.

Hospitalizations related to back pain were more frequent among men (52.7%). The percentage was highest in the 18 to 44 and 45 to 64 age groups and totaled 60.2% and 57.1% respectively of those hospitalized. In the 18 to 44 age group, hospitalizations were more frequent among men for all four categories associated with back pain.

Conditions associated with neck pain resulted in 227,000 hospitalizations and 1.4 million patient days. Slightly more than half of hospitalizations (52.0%) were men. Average length of stay was identical for men and women (6.3 days). For neck injury, however, length of stay was substantially higher among men (8.7 days) than women (6.7 days). In each of the remaining three neck pain categories, average length of stay was higher among women.

Table 1: (part 1) Average Annual Hospitalizations Resulting from Selected Musculoskeletal Conditions: United States 1985-1988, by Age and Gender[1]

	Gender	Total	Less than 18	18-44	45-64	65 & over	Hospital Days[2]	Average Length of Stay
Back pain								
Back Disorders, excluding disk		289,000	3,000	108,000	97,000	81,000	1,871,000	6.5
	Male	135,000	*	57,000	47,000	29,000	808,000	6.0
	Female	154,000	*	51,000	50,000	52,000	1,063,000	6.9
Disk disorders, displacement		334,000	2,000	187,000	114,000	32,000	2,247,000	6.7
	Male	200,000	1,000	120,000	65,000	13,000	1,253,000	6.3
	Female	134,000	1,000	67,000	48,000	18,000	994,000	7.4
Other disk disorders		80,000	*	32,000	31,000	16,000	569,000	7.2
	Male	39,000	*	19,000	15,000	5,000	264,000	6.7
	Female	40,000	*	14,000	16,000	11,000	305,000	7.6
Back injury		184,000	6,000	93,000	46,000	39,000	1,273,000	6.9
	Male	93,000	3,000	57,000	22,000	12,000	614,000	6.6
	Female	90,000	3,000	36,000	24,000	27,000	658,000	7.3
Total back		886,000	11,000	420,000	288,000	168,000	5,960,000	6.7
	Male	467,000	5,000	253,000	149,000	59,000	2,940,000	6.3
	Female	418,000	5,000	168,000	138,000	108,000	3,020,000	7.2
Neck pain								
Neck disorders excluding disk		90,000	*	33,000	38,000	18,000	518,000	5.8
	Male	46,000	*	17,000	20,000	9,000	255,000	5.5
	Female	44,000	*	16,000	18,000	9,000	263,000	6.0
Disk disorders displacement		42,000	*	22,000	18,000	2,000	234,000	5.6
	Male	22,000	*	12,000	9,000	*	116,000	5.3
	Female	20,000	*	10,000	9,000	*	117,000	6.0
Other disk disorders		24,000	*	11,000	10,000	3,000	127,000	5.3
	Male	13,000	*	7,000	4,000	*	53,000	4.2
	Female	11,000	*	5,000	5,000	*	74,000	6.6
Neck injury		71,000	6,000	46,000	14,000	5,000	549,000	7.7
	Male	37,000	4,000	24,000	7,000	3,000	319,000	8.7
	Female	34,000	2,000	23,000	7,000	3,000	230,000	6.7
Total neck		227,000	7,000	113,000	80,000	28,000	1,428,000	6.3
	Male	118,000	4,000	60,000	40,000	14,000	743,000	6.3
	Female	109,000	2,000	54,000	39,000	14,000	685,000	6.3

[1]First listed diagnosis for inpatients discharged from short-stay hospitals.
[2]Annual Average
**Estimate does not meet standards of reliability or precision.*

Source: *National Center for Health Statistics, National Hospital Discharge Survey, 1985-88.*

Table 1: (part 2) Average Annual Hospitalizations Resulting from Selected Musculoskeletal Conditions: United States 1985-1988, by Age and Gender[1]

	Gender	Total	Less than 18	18-44	45-64	65 & over	Hospital Days[2]	Average Length of Stay
Arthritis								
Rheumatoid arthritis and other inflammatory polyarthropathies		57,000	*	8,000	24,000	23,000	438,000	7.7
	Male	13,000	*	2,000	5,000	5,000	100,000	7.5
	Female	44,000	*	6,000	19,000	18,000	338,000	7.7
Osteoarthrosis and allied disorders		206,000	*	11,000	57,000	138,000	2,083,000	10.1
	Male	77,000	*	6,000	23,000	47,000	727,000	9.4
	Female	129,000	*	4,000	33,000	91,000	1,356,000	10.5
Other & unspecified arthropathies and related disorders		49,000	5,000	15,000	14,000	15,000	667,000	8.3
	Male	25,000	3,000	10,000	6,000	4,000	264,000	7.7
	Female	24,000	3,000	4,000	7,000	10,000	403,000	8.7
Total arthritis and related disorders		312,000	8,000	34,000	95,000	175,000	3,188,000	9.3
	Male	116,000	5,000	19,000	35,000	58,000	1,091,000	8.7
	Female	197,000	3,000	15,000	60,000	118,000	2,097,000	9.6
Congenital malformations of the musculoskeletal system		54,000	27,000	15,000	9,000	4,000	321,000	6.0
	Male	27,000	15,000	8,000	4,000	*	169,000	6.2
	Female	27,000	12,000	7,000	5,000	3,000	152,000	5.7
Neoplasms								
Malignant neoplasms of bone and connective tissue		90,000	6,000	8,000	28,000	48,000	921,000	10.3
	Male	43,000	5,000	4,000	11,000	23,000	425,000	9.9
	Female	47,000	*	4,000	16,000	25,000	496,000	10.7
Benign neoplasms of bone and connective tissue		29,000	5,000	11,000	8,000	4,000	90,000	3.2
	Male	13,000	4,000	5,000	3,000	*	38,000	3.0
	Female	16,000	*	7,000	5,000	*	52,000	3.3
Total neoplasms		118,000	11,000	19,000	36,000	52,000	1,011,000	8.6
	Male	56,000	8,000	9,000	14,000	24,000	463,000	8.3
	Female	62,000	3,000	10,000	22,000	28,000	548,000	8.8
Total, above musculoskeletal conditions		1,598,000	64,000	601,000	508,000	427,000		

[1] *First listed diagnosis for inpatients discharged from short-stay hospitals.*
[2] *Annual Average*
* *Estimate does not meet standards of reliability or precision.*

Source: *National Center for Health Statistics, National Hospital Discharge Survey, 1985-88.*

Arthritic conditions resulted in 312,000 hospitalizations per year and were second only to back pain in frequency among musculoskeletal conditions. Almost two-thirds of hospitalizations related to arthritis (66.0%) were a result of osteoarthrosis and allied disorders; 18.3% were a result of rheumatoid arthritis and other inflammatory polyarthropathies. In contrast to back and neck pain, a large majority of those hospitalized because of arthritis were women. Women accounted for 63.1% of those hospitalized as a result of arthritis, including 77.2% hospitalized because of rheumatoid arthritis and 62.6% because of osteoarthritis. Average lengths of stay were also higher for women (9.6 days) than for men (8.7 days) and were higher in each of the three arthritis categories. Osteoarthrosis and allied disorders had the longest length of stay (10.1 days).

Congenital malformations resulted in approximately 54,000 hospitalizations per year and were virtually evenly divided between men and women.

An estimated 118,000 hospitalizations resulted from neoplasms of bone and connective tissue. Malignant neoplasms accounted for 76.3% of hospitalizations and 91.1% of the patient days in this category. The majority of hospitalizations for neoplasms (52.5%) occurred among women, and lengths of stay were also longer among women for both malignant and benign neoplasms.

Physician visits

The five conditions discussed above also result in large numbers of physician visits. Table 2 indicates the estimated number of visits to physicians in office-based practice in the United States for each of these conditions.

Almost 14.3 million office visits resulted from conditions associated with back pain. The largest percent, 42.4%, were the result of non-disk related disorders, while 41.2% resulted from back injury and 16.5% from disk disorders.

Although slightly over half of visits for conditions associated with back pain are made by men (50.7%), the distribution of visits by gender varies by the type of back condition and age group. Men, for instance, account for 57.1% of visits for disk disorders and 55.1% of visits relating to back injury. The disparity by gender is greatest in the 18 to 44 age group where men account for 60.5% of visits related to disk disorders and 57.8% attributable to back injury.

Table 2: (part 1) Visits to Physicians in Office-based Practice for Selected Musculoskeletal Conditions: United States 1985, by Age and Gender[1]

(thousands)

	Gender	Total	Less than 18	18-44	45-64	65 & over
Back pain						
Back disorders, excluding disk		6,055	*	2,412	2,011	1,460
	Male	2,653	*	1,033	949	577
	Female	3,402	*	1,379	1,062	884
Disk disorders		2,357	*	1,171	891	275
	Male	1,347	*	709	519	*
	Female	1,010	*	463	372	*
Back injury		5,882	265	3,662	1,390	565
	Male	3,240	*	2,118	733	*
	Female	2,642	*	1,544	657	328
Total back		14,294	458	7,245	4,290	2,301
	Male	7,240	266	3,860	2,200	915
	Female	7,054	*	3,386	2,090	1,386
Neck pain						
Neck disorders excluding disk		2,005	*	726	739	483
	Male	899	*	252	350	259
	Female	1,105	*	474	389	*
Disk disorders		323	*	*	*	*
	Male	*	*	*	*	*
	Female	*	*	*	*	*
Neck injury		2,892	*	1,823	818	*
	Male	1,045	*	650	291	*
	Female	1,848	*	1,173	527	*
Total neck		5,221	*	2,673	1,742	611
	Male	2,081	*	962	716	311
	Female	3,140	*	1,711	1,026	300

[1]Principal diagnosis associated with patient's reason for visit.
**Estimate does not meet standards of reliability or precision.*

Source: National Ambulatory Medical Care Survey, 1985.

Table 2: (part 2) Visits to Physicians in Office-based Practice for Selected Musculoskeletal Conditions: United States 1985, by Age and Gender[1]

	Gender	Total	Less than 18	18-44 (thousands)	45-64	65 & over
Arthritis						
Rheumatoid arthritis and other inflammatory polyarthropathies		2,098	*	685	1,051	1,084
	Male	762	*	*	297	320
	Female	2,146	*	547	754	764
Osteoarthrosis and allied disorders		5,222	*	480	1,815	3,170
	Male	1,859	*	*	756	907
	Female	3,663	*	324	1,060	2,262
Other and unspecified arthropathies and related disorders		4,280	*	813	1,502	1,891
	Male	1,521	*	354	508	645
	Female	2,758	*	459	994	1,246
Total arthritis and related disorders		12,710	*	1,978	4,367	6,145
	Male	4,141	*	648	1,561	1,872
	Female	8,569	*	1,329	2,807	4,273
Congenital malformations of the musculoskeletal system		628	269	*	*	*
	Male	339	*	*	*	*
	Female	289	*	*	*	*
Neoplasms		818	*	289	279	*
	Male	291	*	*	*	*
	Female	527	*	*	*	*

[1] *Principal diagnosis associated with patient's reason for visit.*
**Estimate does not meet standards of reliability or precision.*

Source: National Ambulatory Medical Care Survey, 1985.

Approximately 5.2 million office visits were made for conditions related to neck pain. In contrast to back pain, a sizeable majority of visits were made by women (60.1%), and women accounted for a majority of visits in each neck-pain category.

More than 12.7 million visits were made to office-based physicians for arthritic conditions. A plurality, 41.1%, resulted from osteoarthrosis and allied disorders; 22.9% resulted from rheumatoid arthritis and other inflammatory polyarthropathies. More than two-thirds of visits (67.5%) were made by women including 70.1% of osteoarthrosis visits and 73.8% of rheumatoid arthritis visits.

Compared with the other musculoskeletal categories, significantly fewer visits resulted from congenital malformation of the musculoskeletal system (628,000) and neoplasms of bone and connective tissue (818,000). In contrast to inpatient care, in which a large majority of hospitalizations and patient days resulted from malignant neoplasms, 77.4% of office visits related to neoplasms were a result of benign neoplasms (not shown).

For the conditions listed in Table 3, disk disorders and arthritis result in the largest number of return visits per condition (3.3 return visits per new-problem office visit). Congenital malformations necessitated the smallest number of return visits per condition.

Table 3: Visits to Physicians in Office-based Practice: United States 1985, Type of Visit for Selected Musculoskeletal Conditions[1]

(thousands)

	Total visits	New problem visits	Referred new problem visits	Return visits/ new problem visits	New problem percent referred
Back pain	**14,294**	**5,272**	**1,145**	**1.7**	**21.7**
Disc disorders	**2,357**	**552**	***215**	**3.3**	***38.9**
Neck Pain	**5,221**	**1,991**	**459**	**1.6**	**23.1**
Arthritis	**12,710**	**2,926**	**579**	**3.3**	**19.8**
Neoplasms	**818**	**299**	***94**	**1.7**	***31.4**
Congenital malformations	**628**	**287**	***113**	**1.2**	***39.4**

[1]Principal diagnosis associated with patient's reasons for visit.
**Estimate does not meet standards of reliability or precision.*

Source: *National Ambulatory Medical Care Survey, 1985.*

Examining new-problem office visits indicates that congenital malformations and disk disorders were the conditions most likely to be referred (39.4% and 38.9% respectively). Arthritis cases were only half as likely to be referred, with 19.8% of new-problem office visits being referred.

Physician visits by specialty

Figures 1-4 show new-problem office visits by specialty without referral and new-problem office visits with referral for four conditions: back pain, neck pain, disk disorders and arthritis. For new-problem back pain cases in which the patient was not referred, 60.6% of patients were treated by primary care physicians: general/family practitioners, 47.4% and internists, 13.2% (Figure 1). Eighteen percent were treated by orthopaedic surgeons; 21.4% by other specialties. When a referral was made, 43.2% were referred to orthopaedic surgeons; 11.7% to neurosurgeons; 9.8% to neurologists.

Figure 1: Percent Distribution of New Problem Office Visits by Physician Specialty Back Pain, All Sites

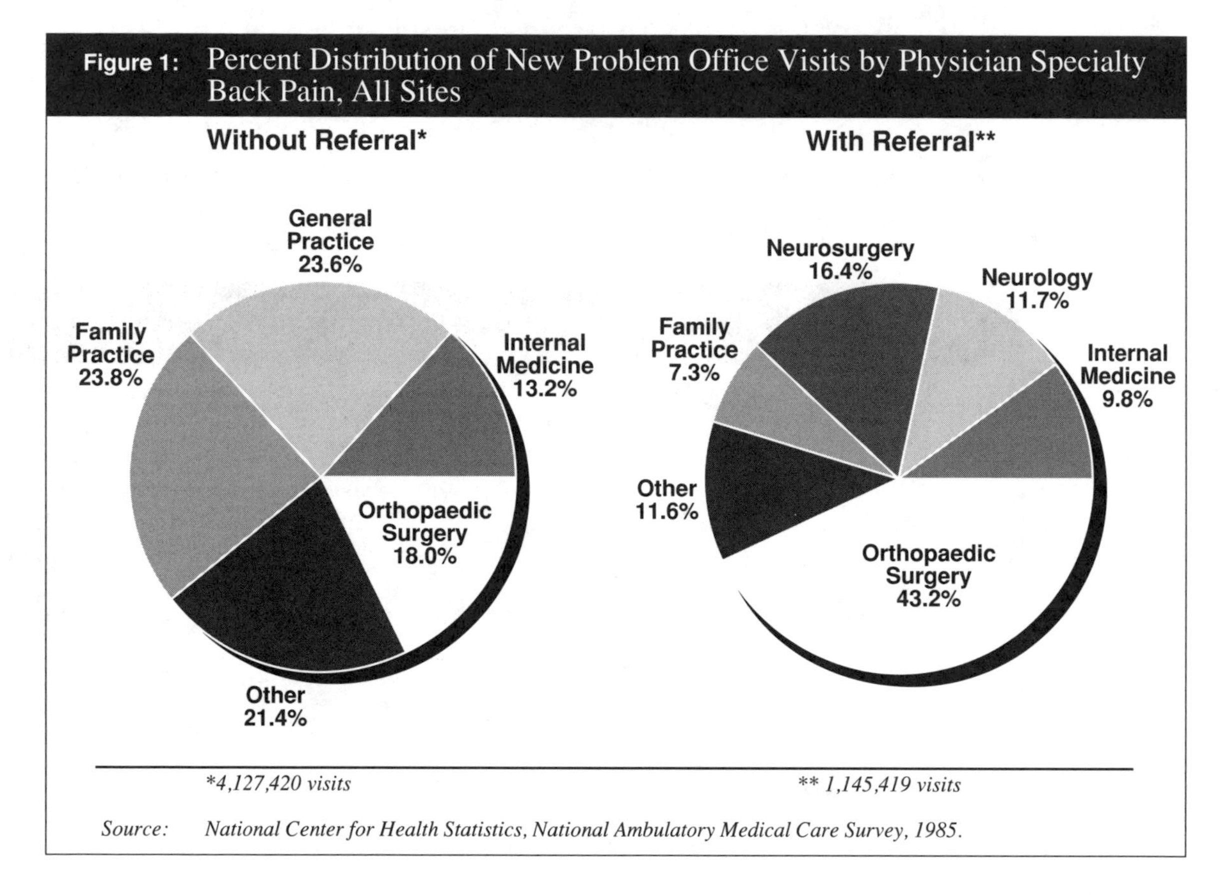

*4,127,420 visits

** 1,145,419 visits

Source: *National Center for Health Statistics, National Ambulatory Medical Care Survey, 1985.*

For neck pain without referral, 61.8% of patients making a new-problem visit were treated by primary care physicians; 20.7% by orthopaedic surgeons (Figure 2). With referral, the largest percent, 29.9% were referred to orthopaedic surgeons; 22.9% to neurosurgeons; 17.7% to neurologists.

In relation to disk disorders, the majority of new-problem office visits without referral were made to surgical specialties (Figure 3). More than two-thirds of these visits were made either to orthopaedic surgeons (41.2%) or to neurosurgeons (27.3%). A comparatively small percentage of visits was made to general/family practitioners (23.4%) or internists (3.6%). When a referral was made, 34.7% were made to orthopaedic surgeons, 27.1% to neurosurgeons and 12.9% to rheumatologists.

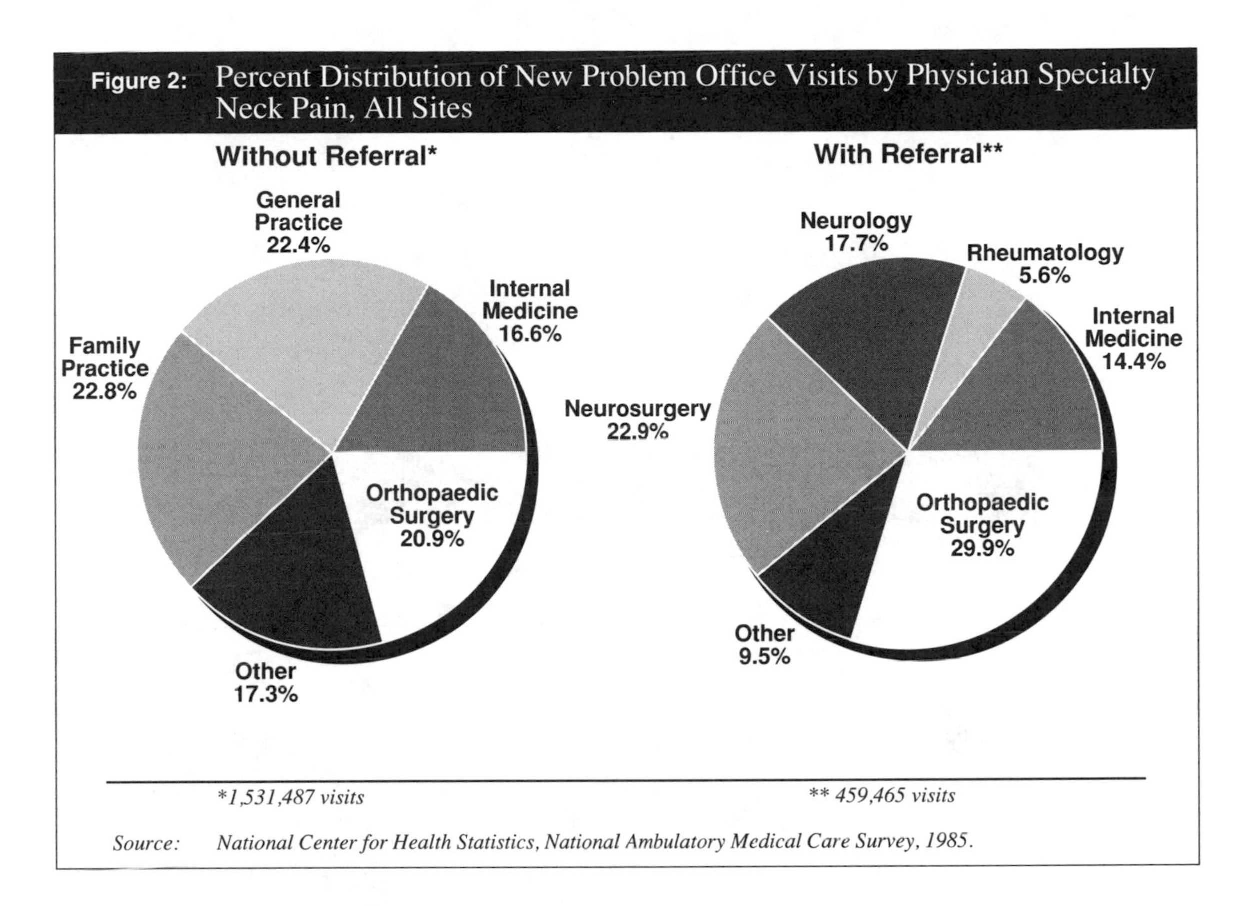

Figure 2: Percent Distribution of New Problem Office Visits by Physician Specialty Neck Pain, All Sites

1,531,487 visits* * 459,465 visits*

Source: *National Center for Health Statistics, National Ambulatory Medical Care Survey, 1985.*

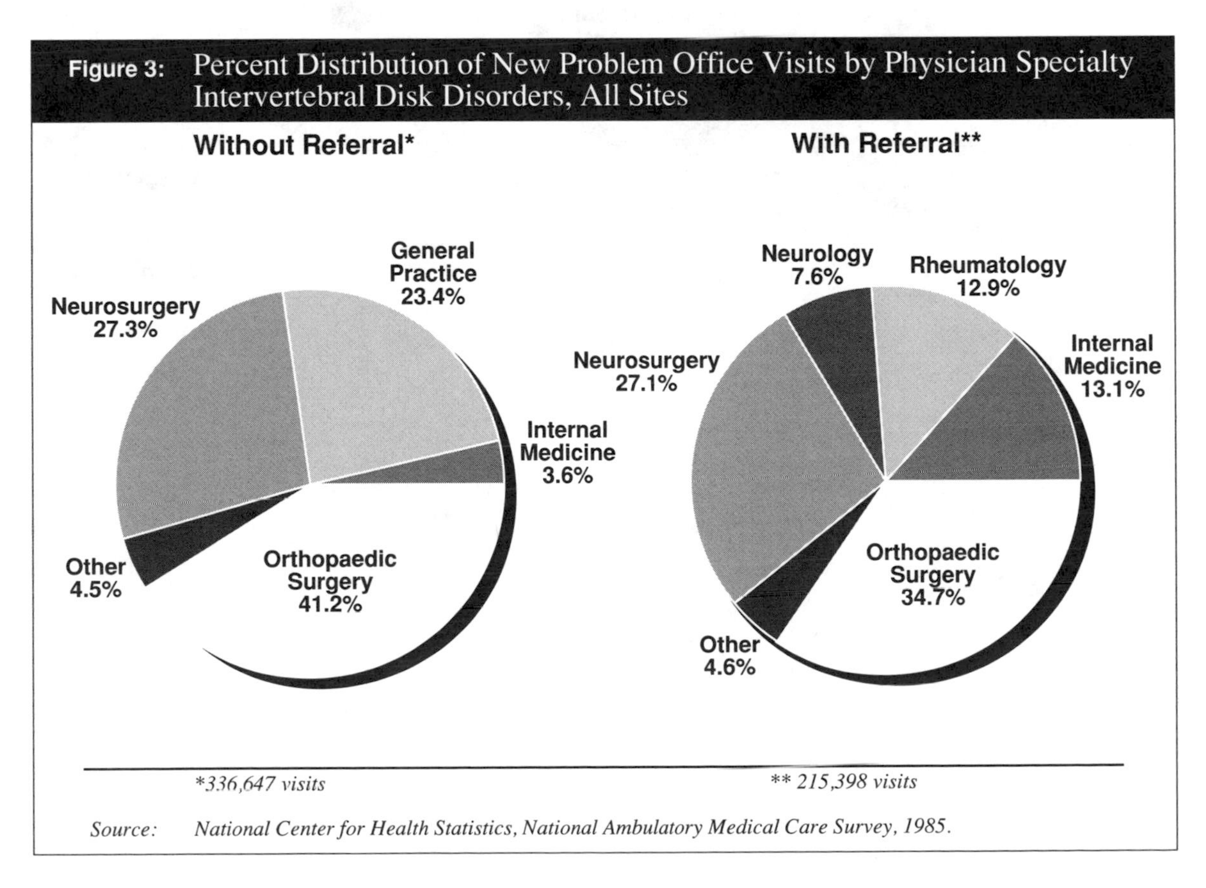

Figure 3: Percent Distribution of New Problem Office Visits by Physician Specialty Intervertebral Disk Disorders, All Sites

336,647 visits* * 215,398 visits*

Source: *National Center for Health Statistics, National Ambulatory Medical Care Survey, 1985.*

Of the four musculoskeletal conditions examined, new-problem office visits without referral for arthritis were the most likely to be made to primary care physicians (68.4%) (Figure 4). Relatively small percentages of unreferred visits were made to other specialists: orthopaedic surgeons (14.9%), general surgeons (3.8%), rheumatologists (3.5%). When a referral was made, most were referred either to orthopaedic surgeons (49.7%) or to rheumatologists (20.8%).

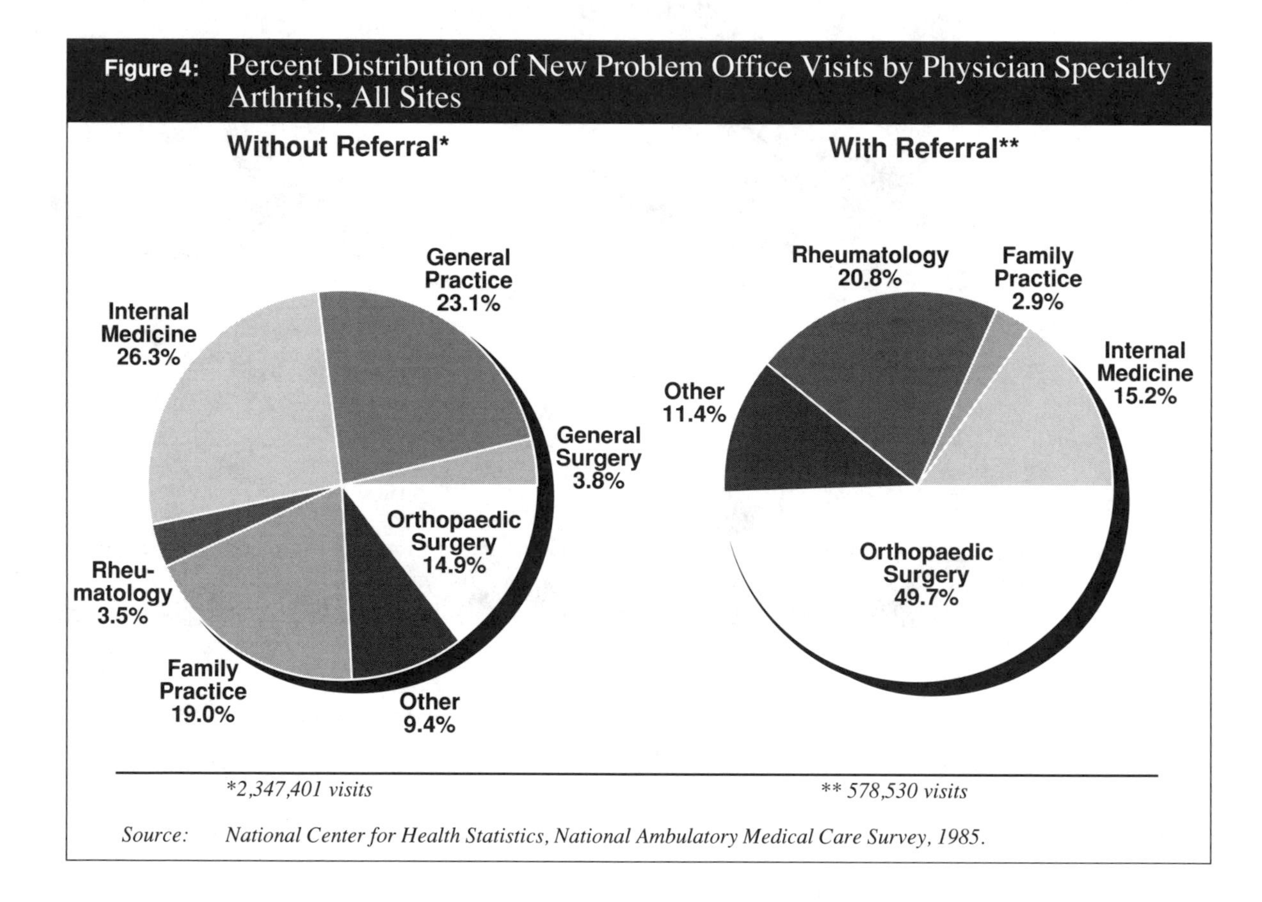

Figure 4: Percent Distribution of New Problem Office Visits by Physician Specialty Arthritis, All Sites

2,347,401 visits* * 578,530 visits*

Source: National Center for Health Statistics, National Ambulatory Medical Care Survey, 1985.

References

National Center for Health Statistics, National Ambulatory Medical Care Survey, 1985 (Data Tape).

National Center for Health Statistics, National Hospital Discharge Survey, 1985-1988 (Data Tape).

Section 3

Musculoskeletal Injuries

chapter 1

Frequency of Occurrence

The average annual number of persons injured in the United States from 1985 through 1988 was 61.1 million (Table 1) or 25.8/100* persons in the civilian noninstitutionalized population (Table 2). Injuries, reported each year in the National Health Interview Survey, include those that result in either medical attention or at least one-half day of restricted activity. Of injuries reported, approximately 32.8 million (53.6%) were injuries to the musculoskeletal system, an incidence rate of 13.8/100 persons.

Table 1: Average Annual Number of Episodes of Persons Injured by Type of Injury: United States, 1985-88

(Thousands)

	Total	Male	Female	Less than 18 years	18-44 years	45-64 years	65 years & over
Fractures	6,155	3,361	2,794	2,090	2,387	1,075	604
Neck and trunk	857	490	367	*51	405	194	*207
Humerus, radius and ulna	759	322	436	374	*161	*127	*97
Femur	*159	*23	*136	-	*13	*15	*131
Tibia, fibula and ankle	492	258	234	215	*144	*121	*11
Other limbs	2,958	1,830	1,127	1,274	1,197	392	*95
Dislocations and sprains	14,667	7,860	6,807	3,451	8,196	2,092	928
Crushing injury	289	*192	*96	*66	*146	*61	*15
Open wound	9,480	6,181	3,299	3,027	4,778	1,073	601
Other injury	2,189	1,171	1,018	480	1,018	447	244
Total musculoskeletal injuries	32,780	18,765	14,014	9,112	16,525	4,748	2,392
Total injuries	61,137	33,466	27,671	19,334	28,461	7,781	5,561

**Estimate does not meet standards of reliability or precision.*

Source: *National Center for Health Statistics, National Health Interview Survey. Data tapes, 1985-1988.*

* Data indicate episodes of persons injured.

Table 2: Average Annual Number of Episodes of Persons Injured by Type of Injury: United States, 1985-88

(Rate per 100 persons)

	All injuries	Musculoskeletal injuries	Dislocations and sprains	Fractures
Total	**25.8**	**13.8**	**6.2**	**2.6**
Male	**29.1**	**16.3**	**6.9**	**2.9**
Female	**22.6**	**11.4**	**5.6**	**2.3**
Less than 18 years	**30.6**	**14.4**	**5.5**	**2.1**
18-44 years	**28.1**	**16.3**	**8.1**	**2.4**
45-64 years	**17.3**	**10.6**	**4.7**	**2.4**
65 years & over	**20.0**	**8.6**	**3.3**	**2.2**
65-74 years	**17.8**	**8.0**	**3.3**	**2.0**
75-84 years	**20.6**	**8.6**	**3.4**	***1.8**
85 years & over	**35.4**	**13.7**	***3.2**	***4.9**

**Estimate does not meet standards of reliability or precision.*

Source: National Center for Health Statistics, National Health Interview Survey. Data tapes, 1985-1988.

Distribution by category

The distribution of musculoskeletal injuries by major category is indicated in Figure 1. Dislocations and sprains accounted for the largest musculoskeletal injury category (44.7%) followed by open wounds (28.9%) and fractures (18.8%). It should be noted that open wounds and other injuries, such as contusions, have not universally been considered injuries of the musculoskeletal system. These injuries, however, frequently involve the connective tissue of the musculoskeletal system, most notably ligaments, tendons and muscles, and are often treated by physicians specializing in musculoskeletal care.

For musculoskeletal injuries, the annual rate of persons injured was greater among males than females, 16.3 vs. 11.4/100 persons, respectively (Table 2). Incidence rates were also higher for men in both musculoskeletal subcategories. Dislocations and sprains occurred at a rate of 6.9/100 among men compared with 5.6/100 among women. For fractures, the incidence rates were 2.9/100 among men and 2.3/100 among women.

The average annual rate of episodes of persons injured by age and

Figure 1: Average Annual Number of Musculoskeletal Injuries by Type of Injury: United States, 1985-88

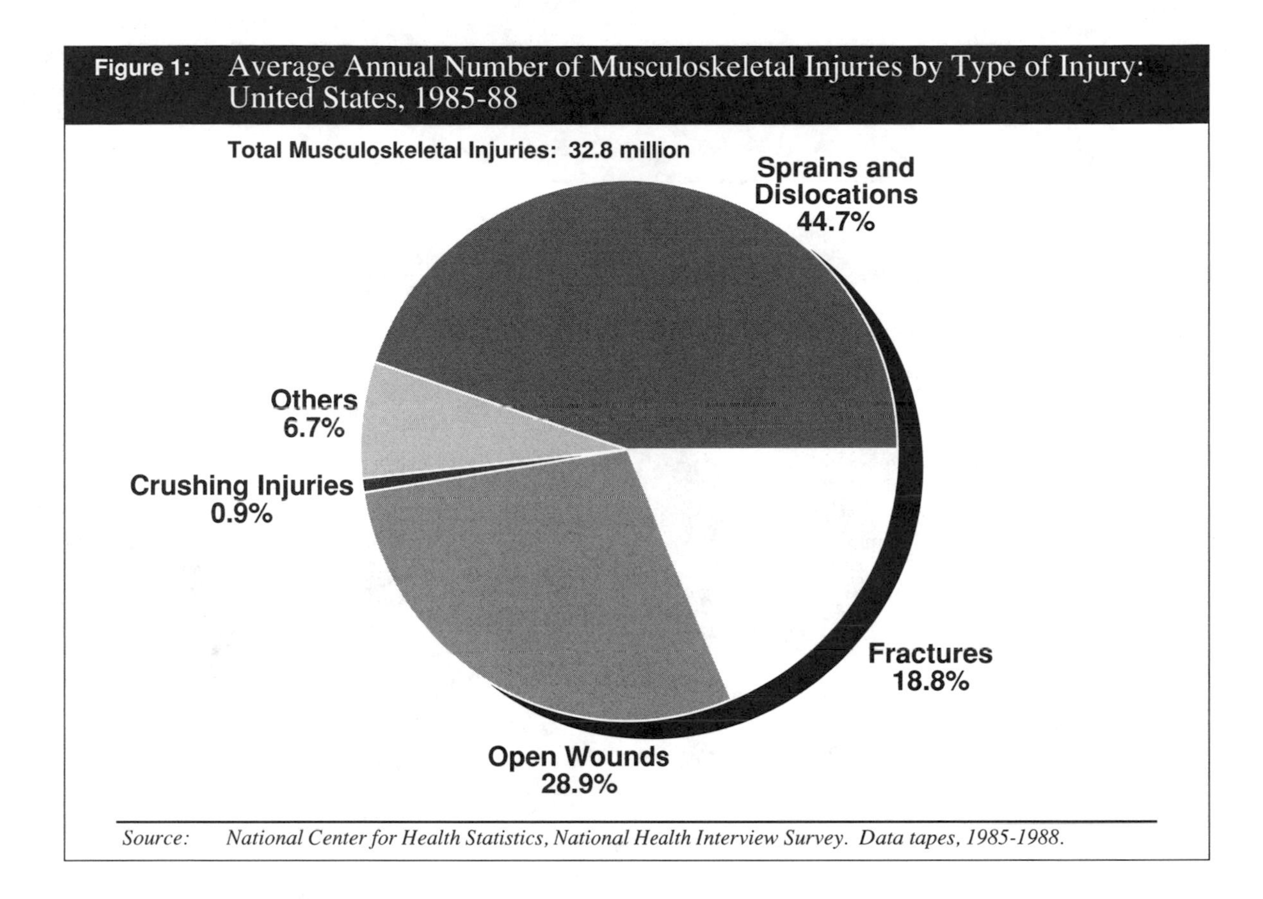

Source: *National Center for Health Statistics, National Health Interview Survey. Data tapes, 1985-1988.*

gender for three categories: all musculoskeletal injuries; dislocations and sprains, and fractures are indicated in Figures 2-4, respectively.

The data show a consistent and substantial variation by age and gender for all three categories. Rates are higher among men than women in the less-than-18 and 18 to 44 age groups. These rates decline successively and become lower than rates among women for groups 65 years of age and older. Rates among women, in contrast, are more consistent over the life cycle.

Distribution by age and gender

For all musculoskeletal injuries considered together (Figure 2), rates among males reach a maximum (20.7/100) in the 18 to 44 age group and fall consistently to 4.7/100 persons among those age 75 to 84 years of age before increasing. Rates among women were more stable and follow a U-shaped pattern, reaching a minimum rate of 8.9/100 persons in the 65 to 74 age group.

The distribution of rates for dislocations and sprains by gender and age group was similar to those for all musculoskeletal injuries, with rates

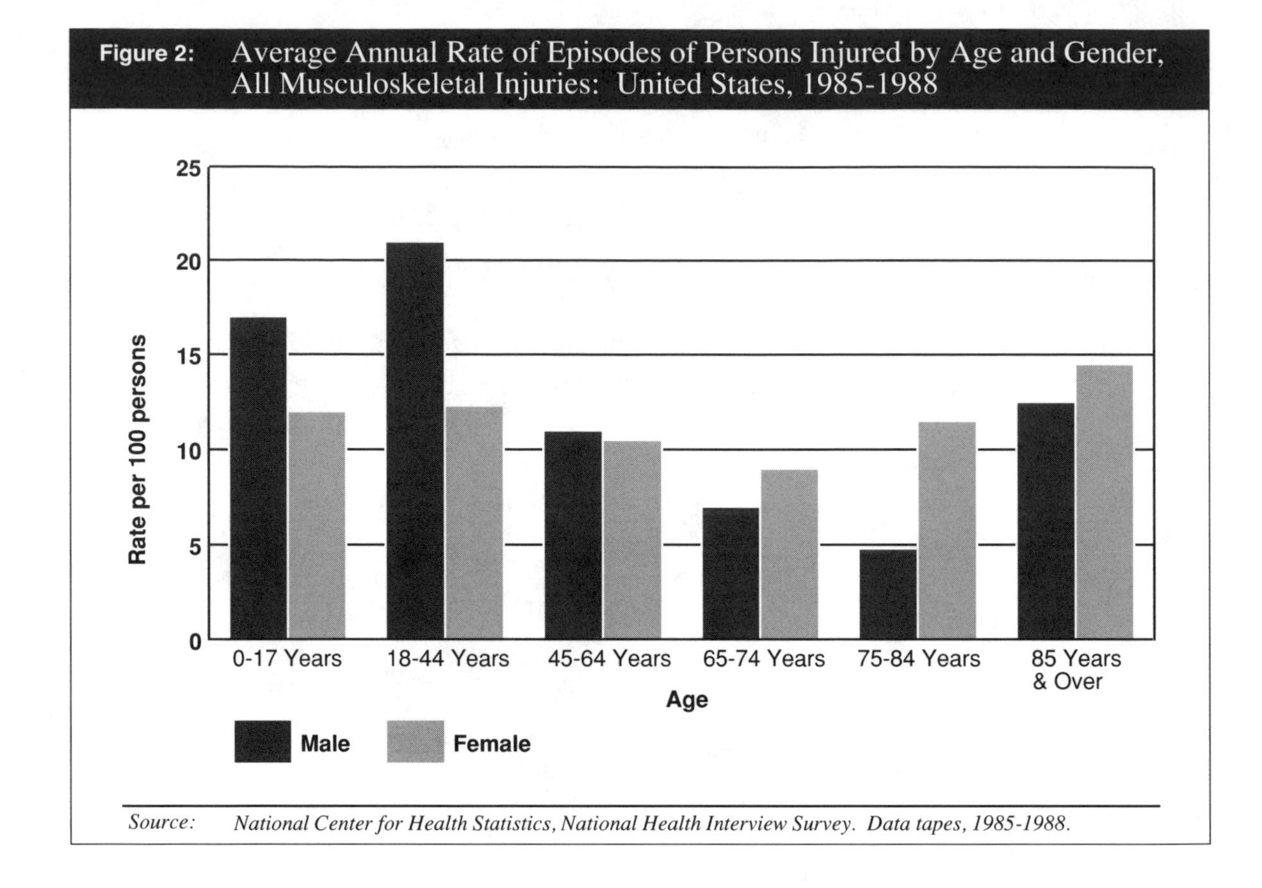

Figure 2: Average Annual Rate of Episodes of Persons Injured by Age and Gender, All Musculoskeletal Injuries: United States, 1985-1988

Source: *National Center for Health Statistics, National Health Interview Survey. Data tapes, 1985-1988.*

for both men and women reaching their maximum in the 18 to 44 age group (Figure 3).

Although fractures exhibit a general pattern similar to that of all musculoskeletal injuries as well as to that of dislocations and sprains, especially for males, there are two major differences (Figure 4). First, fracture rates among women exceed those of men: by the time they enter the 45 to 64 age group. In this age group, fractures occur among women at a rate of 3.0/100 persons compared with 1.8/100 persons among men. For all musculoskeletal injuries and for sprains and dislocations, there were virtually no gender differences in rates among this age group. Second, fracture rates among women increase dramatically in the 85-and-older age group which, in all likelihood, reflects the high incidence of osteoporosis-related fractures among this group (see section on osteoporosis and hip fracture). This increase is substantial enough that the entire increase in rates (11.1/100 to 14.5/100 persons) for all musculoskeletal injuries considered together for women from the 75 to the 84 age group to the 85-and-older age groups (Figure 2) can be attributed to the increase in fracture rates (2.5/100 to 6.4/100 persons).

Figure 3: Average Annual Rate of Episodes of Persons Injured by Age and Gender, Dislocations and Sprains: United States, 1985-1988

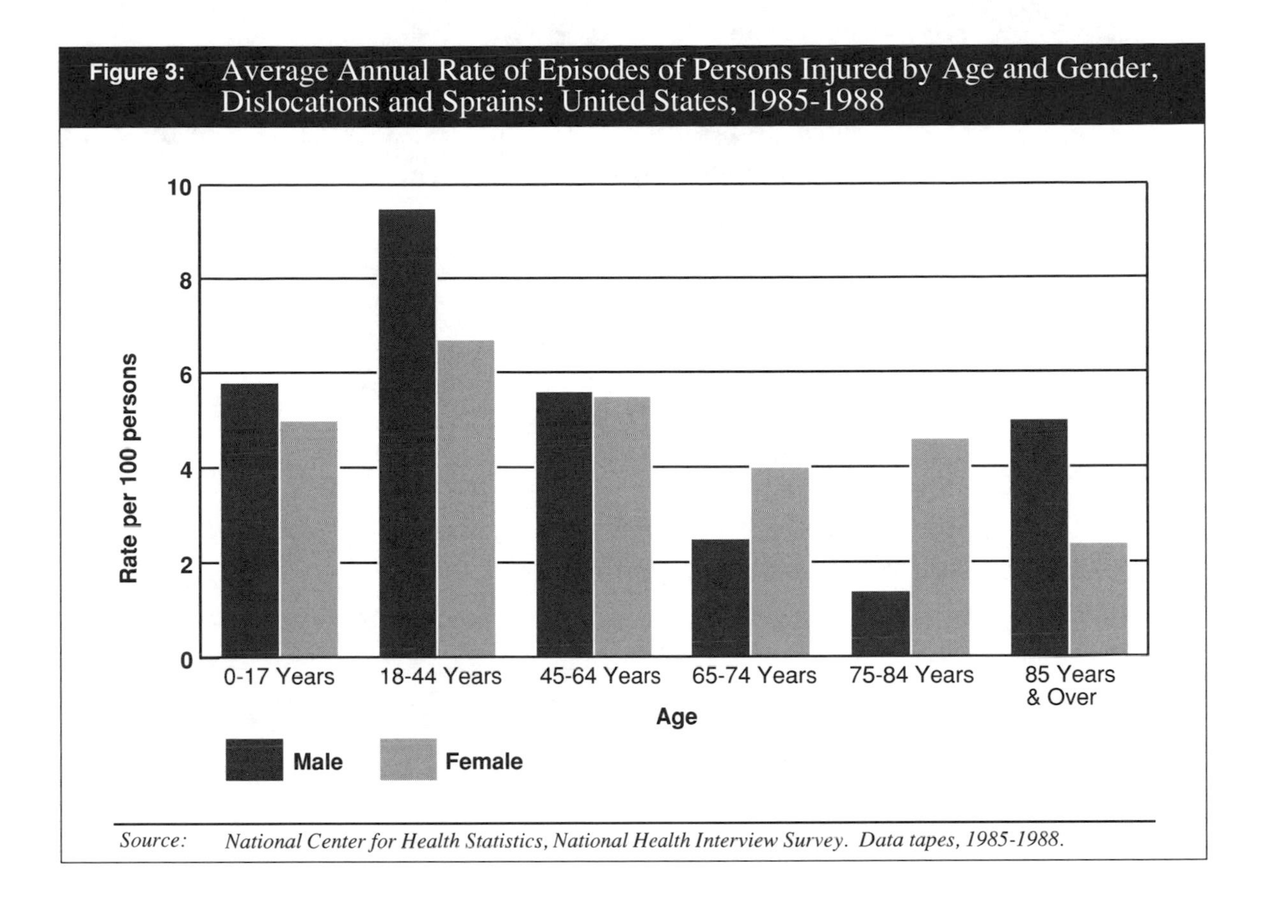

Source: *National Center for Health Statistics, National Health Interview Survey. Data tapes, 1985-1988.*

Figure 4: Average Annual Rate of Episodes of Persons Injured by Age and Gender, Fractures: United States, 1985-1988

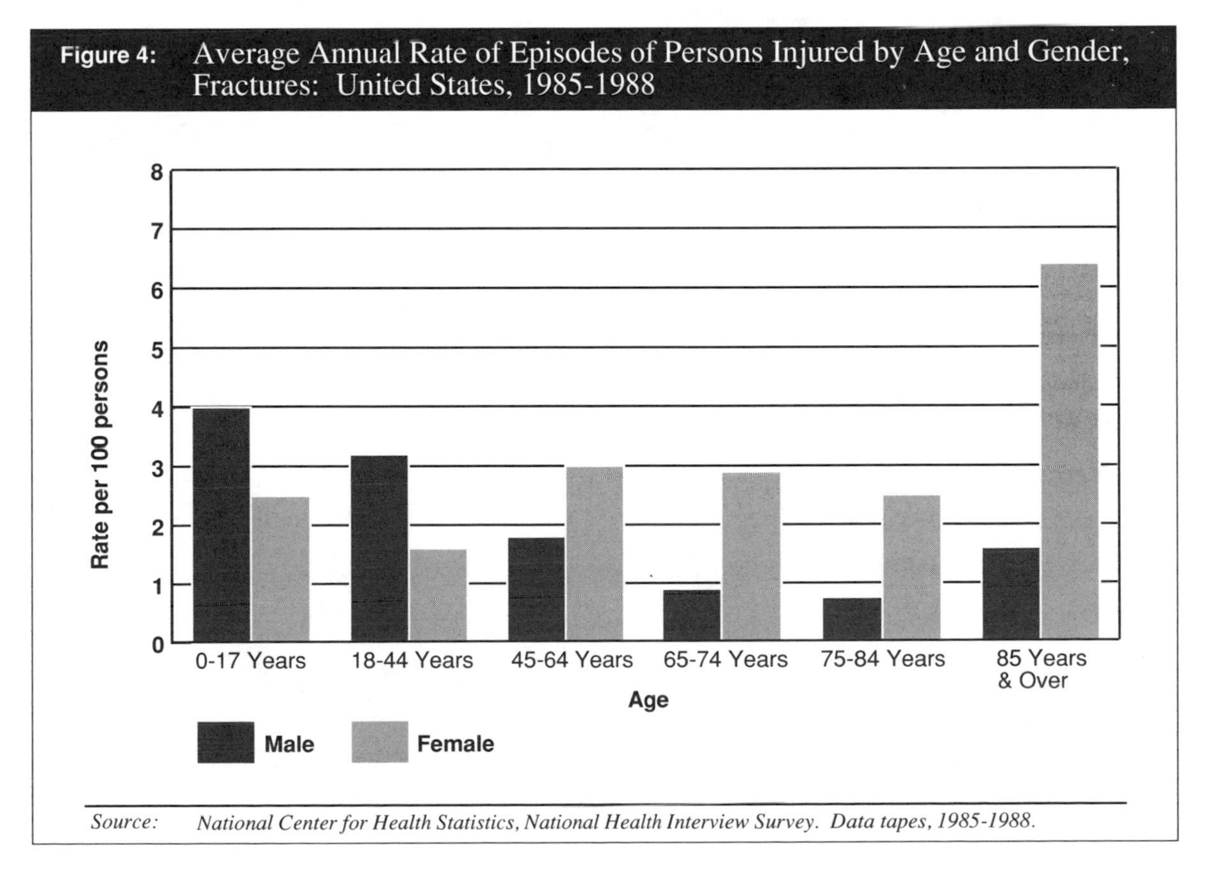

Source: *National Center for Health Statistics, National Health Interview Survey. Data tapes, 1985-1988.*

References

National Center for Health Statistics, National Health Interview Survey, 1985-1988 (Data Tapes).

chapter 2

Place of Occurrence

Figures 1 through 4 show the percent distribution of the known place of occurrence for all injuries; all musculoskeletal injuries; fractures; and dislocations and sprains, respectively. These distinctions are based on data from the 1985-88 National Health Interview Surveys.

The percent distribution of all injuries (Figure 1) and of musculoskeletal injuries (Figure 2) by place of occurrence were similar. Compared with all injuries, however, musculoskeletal injuries were somewhat more likely to have occurred in industrial settings (16.1% vs. 14.3%) and places of recreation (9.8% vs. 8.5%). Musculoskeletal injuries, on the other hand, were less likely to have occurred inside the home (20.2% vs. 22.4%) or on streets and highways (12.0% vs. 13.8%) than were all injuries.

Figure 1: Percent Distribution of Place of Occurence, All Injuries United States, 1985-1988

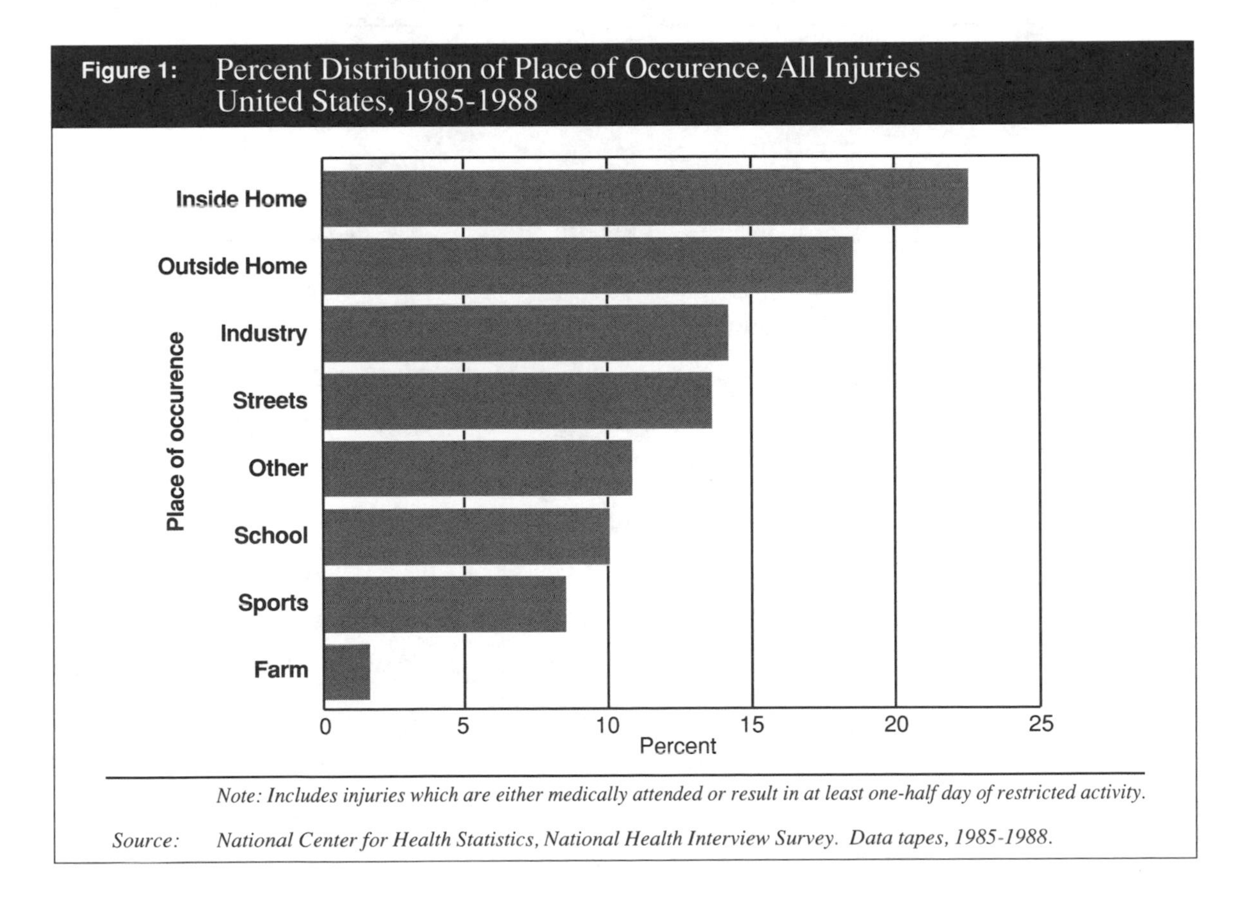

Note: Includes injuries which are either medically attended or result in at least one-half day of restricted activity.

Source: *National Center for Health Statistics, National Health Interview Survey. Data tapes, 1985-1988.*

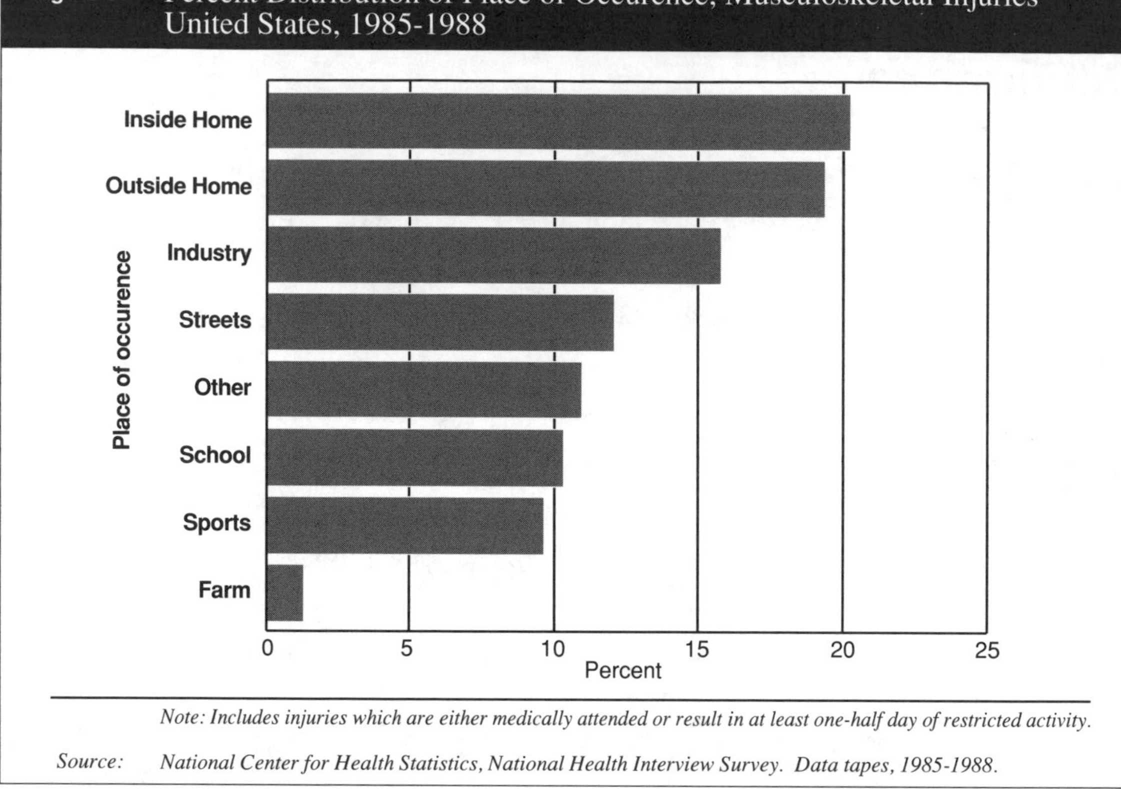

Figure 2: Percent Distribution of Place of Occurence, Musculoskeletal Injuries United States, 1985-1988

Note: Includes injuries which are either medically attended or result in at least one-half day of restricted activity.

Source: National Center for Health Statistics, National Health Interview Survey. Data tapes, 1985-1988.

All musculoskeletal injuries considered together were most frequently reported to have occurred either in (20.2%) or around the home (19.3%). An additional 16.1% occurred in industrial settings. Other places of occurrence included streets and highways (12.0%), schools (10.2%) and places of recreation (9.8%). Comparatively few musculoskeletal injuries occurred on farms (1.5%).

The home was also the most frequent location for fractures (Figure 3). Forty-four percent of fractures occurred at home, with virtually equal percents occurring inside and outside the home (22.1% and 21.9%). Industrial settings were the third most frequent place of occurrence (12.9%). Places of recreation were the next most frequent place of occurrence (11.4%), followed by schools (11.2%) and streets and highways (11.2%).

The distribution of dislocations and sprains was less concentrated (Figure 4). The largest percentage occurred inside the home (16.2%), but industrial settings were the second most frequent source of dislocations and sprains (15.4%). The third most frequent place of occurrence was outside the home (14.8%) followed by streets and highways (14.1%), schools (13.6%) and places of recreations (13.0%).

References

National Center for Health Statistics, National Health Interview Survey, 1985-1988 (Data Tapes).

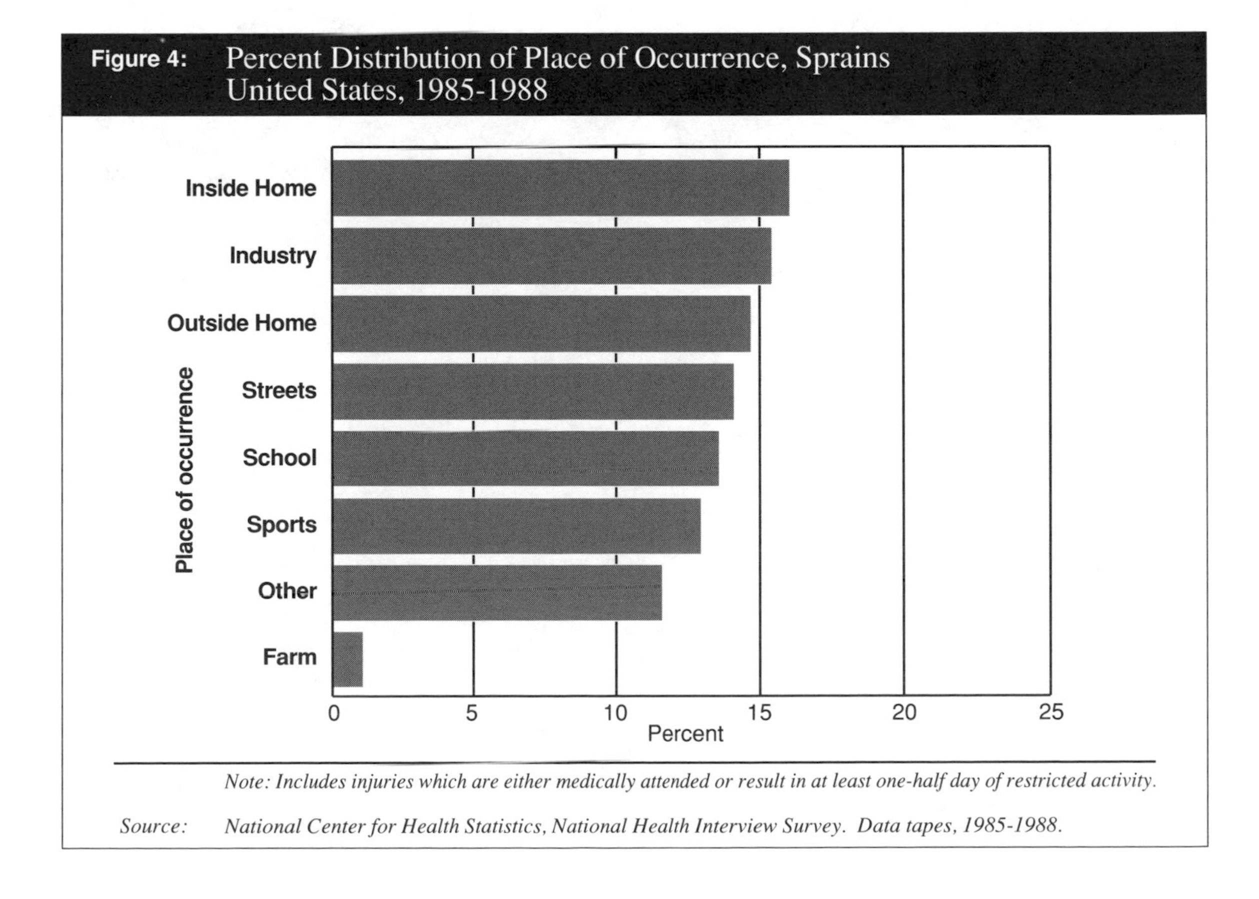

Figure 4: Percent Distribution of Place of Occurrence, Sprains United States, 1985-1988

Note: Includes injuries which are either medically attended or result in at least one-half day of restricted activity.

Source: National Center for Health Statistics, National Health Interview Survey. Data tapes, 1985-1988.

chapter 3

Injuries with Motor-Vehicle Involvement

Motor vehicle accidents are a significant source of injuries, especially to the musculoskeletal system. In 1988, there were approximately 6.9 million police-reported motor vehicle accidents (General Estimates System). These accidents resulted in approximately 3.2 million motor vehicle occupants being killed or injured, with nearly 525,000 of these sustaining a severe or fatal injury. In addition, 187,000 pedestrians or cyclists were injured or killed in vehicular accidents.

The percentages of injuries with reported motor vehicle involvement from the 1985 through 1988 National Health Interview Surveys are indicated in Table 1. Overall, 11.4% of injuries were reported to have resulted from incidents involving motor vehicles. Males reported a larger percentage of injuries with motor vehicle involvement (12.2%) than did females (10.6%).

Table 1: Percentage of Injuries with Motor Vehicle Involvement by Type of Injury and Gender: United States, 1985-88

	Percent with motor vehicle involvement		
Injury	Total	Male	Female
All injuries	11.4	12.2	10.6
Fractures	11.6	13.9	8.9
Neck and trunk	*18.7	*14.3	*24.5
Tibia, fibula and ankle	*22.6	*33.3	*10.8
Humerus, radius and ulna	*8.3	*19.6	-
Other fractures	12.1	*12.3	*11.9
Dislocations and sprains	12.1	11.1	13.4
Open wounds	7.7	7.3	8.3
Other injuries	19.6	*15.9	23.8
All musculoskeletal injuries	10.7	10.2	11.4

*Estimate does not meet standards of reliability or precision.
Note: Includes injuries which are either medically attended or result in at least one-half day of restricted activity.

Source: National Center for Health Statistics, National Health Interview Survey. Data tapes, 1985-1988.

The percentage of reported musculoskeletal injuries with motor vehicle involvement (10.7%) was less than that for all injuries, and the distribution by gender was reversed. For musculoskeletal injuries, the percent attributable to motor vehicle involvement was less among males (10.2%) than among females (11.4%).

Injuries by category

Among the major categories of musculoskeletal injuries, the percentage resulting from motor vehicle involvement was highest for the "other injuries" category (19.6%), followed by dislocations and sprains (12.1%) and fractures (11.6%). The high percent of "other injuries" attributable to motor vehicle involvement is not surprising. This category, although it constitutes a relatively small percentage of musculoskeletal injuries (6.7%), includes a number of types of severe trauma to the musculoskeletal system that would be expected to result from vehicular impact. Included in this category are injuries such as crushing injuries, traumatic amputations, as well as injuries to peripheral nerves of the pelvic girdle and lower limbs.

Although fractures were somewhat more likely to result from motor vehicle accidents (11.6%) than were all musculoskeletal injuries (10.7%), the percentages vary considerably by the site of fracture. Fractures of the lower extremity (excluding the foot) and of the neck and trunk were much more likely to be the result of motor vehicle involvement, 22.6% and 18.7% respectively, than were fractures of the humerus, radius and ulna (8.3%).

Injuries by gender

Men were much more likely to indicate that a fracture resulted from motor vehicle involvement (13.9% of fractures) than were women (8.9%). Men were also more likely to report that fractures of the lower and upper extremities resulted from motor vehicle involvement; women reported that a higher percentage of neck and trunk fractures were the result of motor vehicle accidents.

In contrast to fractures, in the other three categories—dislocations and sprains, open wounds, and other injuries, the percentage indicating motor vehicle involvement was greater among women than men.

References

National Center for Health Statistics, National Health Interview Survey, 1985-1988 (Data Tapes).

National Highway Traffic Safety Administration, General Estimates System 1988. Washington, D.C. 1989.

chapter 4

Occupational Injuries

The workplace is a significant source of injury, occupational illness, including cumulative trauma disorders, and disability. There were an estimated 1.8 million disabling work injuries in the United States in 1990. Of these 10,500 resulted in fatalities and 60,000 in permanent impairment (National Safety Council, "Accident Facts," 1991 edition).

According to the Bureau of Labor Statistics (BLS), in 1988 there were 8.6 recordable cases of occupational injuries or illnesses per 100 full-time workers (Occupational Injuries and Illnesses 1988). An occupational injury is defined by the BLS as any injury which results from a work-related accident or exposure involving a sudden event in the work environment. An occupational illness is any abnormal condition or disorder, other than one resulting from an occupational injury, caused by exposure to factors associated with the employment. Cumulative trauma, or repetitive motion disorders, are included under occupational illness. Approximately 47% of these cases (4.0/100) involved lost workdays and resulted in 76.1 lost workdays per 100 workers for 1988.

Distribution by type

The distribution of occupational injuries and illnesses by type of accident or exposure is indicated in Table 1. These data are derived from the Supplemental Data System (SDS), a Federal-State cooperative program administered by the Bureau of Labor Statistics, which compiles occupational injury and illness data from workers' compensation records in 24 states. Almost 72% of injuries resulted from 3 causes: overexertion (31.2%), being struck by or against another object (23.6%), or falling (17.0%) (Department of Labor, Announcement 90-1).

Injuries to the musculoskeletal system account for the majority of occupational injuries or illnesses that result in work loss (Table 2). Sprains and strains were the most frequent type of injury and accounted for 43.0% of cases involving work loss. Other musculoskeletal categories include: fractures, which accounted for 9.6% of cases; dislocations, 2.0%; inflammation or irritation of joints, tendons and muscles, 1.1%; and amputations (0.6%). In addition, other types of injuries, such as contu-

Table 1: Occupational Injuries or Illnesses by Type of Accident or Exposure
Percent Distribution of Cases Involving Work Loss: 1987

	Percent
Overexertion	**31.2**
Struck by or against	**23.6**
Falls	**17.0**
Bodily reaction	**7.3**
Caught in, under, between	**5.7**
Contact with radiation, caustics	**3.3**
Motor vehicle	**3.0**
All other	**8.9**
	100.0

Source: *US Department of Labor, Bureau of Labor Statistics, Injury and Illness Data Available From 1987 Workers Compensation Records, May 1990 Announcement 90-1.*

Table 2: Occupational Injury and Illness, Type of Injury
Percent Distribution of Cases Involving Work Loss by Gender: 1987

		(Percent)	
	Total	Male	Female
Dislocation	**2.0**	**2.1**	**1.7**
Fracture	**9.6**	**10.5**	**7.4**
Inflammation or irritation of joints, tendons or muscles	**1.1**	**0.7**	**2.2**
Sprain and strain	**43.0**	**40.9**	**48.2**
Amputation	**0.6**	**0.7**	**0.2**
Contusion, crushing, bruise	**9.2**	**8.9**	**10.0**
Cut, laceration, puncture	**11.6**	**13.3**	**7.5**
Scratch, abrasion	**2.5**	**3.0**	**1.2**
Multiple injuries	**2.9**	**3.0**	**2.9**
All other	**17.5**	**16.9**	**18.7**
	100.0	**100.0**	**100.0**

Source: *US Department of Labor, Bureau of Labor Statistics, Injury and Illness Data Available From 1987 Workers Compensation Records, May 1990 Announcement 90-1.*

sions and crushing injuries and lacerations, especially when severe, involve the musculoskeletal system.

Distribution by gender

A comparison of the major types of musculoskeletal injuries by gender, reveals that sprains and strains accounted for a plurality among both men and women. They accounted for almost half of occupational injuries and illnesses among women (48.2%) and 40.9% among men. Inflammation or irritation of joints, tendons or muscles also accounted for a higher percentage of injuries among women (2.2%) than men (0.7%). Fractures, on the other hand, accounted for a higher percentage of injuries among men (10.5% vs. 7.4%) than women. The same was true for dislocations (2.1% vs. 1.7%).

Distribution by anatomic site

The distribution of cases involving fractures and sprains and strains by anatomic site is shown in Table 3. Fractures most frequently involve the upper (42.0%) or lower (38.5%) extremities. Fewer involve the trunk (12.9%), and even fewer the neck (0.3%). Among extremity fractures, fractures of the finger were most frequent and accounted for 18.9% of all fractures, followed by foot (11.5%) and toe fractures (10.1%).

In contrast, most sprains and strains affect the trunk (60.6%), with approximately 80% of these involving the back. About 48.3% of all occupational sprains and strains involve the back. An additional 5.8% involve the shoulder.

Nineteen percent of sprains and strains involve the lower extremities, primarily the knee (7.9%) or ankle (7.0%). Upper extremity sites account for 9.1% of sprains and strains and most frequently involve the wrist (3.3%).

SDS data indicate the back is the most frequent anatomic site injured, accounting for 23.9% of all occupational injuries. When a cause was indicated for back injuries, 90.6% resulted from sprains and strains.

Back injuries most frequently involve the lower back. Of over 100,000 work-related claims for back injury reported to the Ohio Bureau of Workers Compensation from 1985 through 1988, 66.5% involved the lower back. General back complaints accounted for 26.6% of complaints, and 6.2% were listed as back injuries with disk involvement. (Ohio Bureau of Workers Compensation, Statistical Research Section).

Cumulative trauma disorders

Cumulative trauma disorders are an additional source of disability associated with the workplace. Included by the BLS in the category of occupational illness, they are alternatively referred to as "repetitive motion disorders", "chronic microtrauma," or "overuse syndrome." These

Table 3: Occupational Injury, Percent Distribution of Cases Involving Work Loss Fractures and Sprains by Anatomic Site: 1987

(Percent)

	Fractures		Sprains and Strains	
Neck		0.3		2.9
Upper extremities:				
Arm	7.6		2.9	
Wrist	8.0		3.3	
Hand	6.1		0.9	
Finger	18.9		1.2	
Multiple	1.1		0.9	
Total, upper extremities*		42.0		9.1
Trunk				
Back	2.9		48.3	
Chest	5.6		1.2	
Hip	2.1		1.0	
Shoulder	1.9		5.8	
Multiple	0.4		2.4	
Total, trunk*		12.9		60.6
Lower extremities				
Thigh	0.3		0.3	
Knee	2.6		7.9	
Lower leg	2.0		0.3	
Multiple leg	0.1		0.1	
Ankle	8.4		7.0	
Foot	11.5		1.6	
Toe	10.1		0.1	
Multiple lower extremity	1.2		0.8	
Total lower extremity*		38.5		19.0
Multiple parts		2.7		8.1
Other**		3.6		0.3
		100.0		100.0

**Includes other sites not listed individually*
***Primarily skull, jaw, face*

Source: *US Department of Labor, Bureau of Labor Statistics, Injury and Illness Data Available From 1987 Workers Compensation Records, May 1990 Announcement 90-1.*

disorders develop in the absence of acute injury and are associated with repeated trauma or occupational tasks characterized by repeated motion, pressure or vibration. They result in a similar syndrome of chronic localized pain and dysfunction with or without objective physical changes. Specific conditions include, among others, carpal tunnel syndrome, epicondylitis, Raynaud's syndrome, and "gamekeepers thumb."

Data about the incidence or prevalence of these disorders in the workplace has significant limitations. First, as indicated by the BLS, measurement problems exist because, in many cases, employers or physicians are unable or unwilling to identify some of these conditions as work-related. A number of work-related disorders may not be reported as such, and, as a result, their incidence in the workplace is underestimated. Second, at least in relation to SDS program data, other conditions, such as noise-induced hearing loss, are also included in the category "disorders associated with repeated trauma." Finally, there are definitional problems. These disorders include a wide range of afflictions and, at present, there is no consistent, universally accepted set of disorders that are classified as cumulative trauma disorders.

Given these caveats, SDS data indicate that 115,400 cases involving disorders associated with repeated trauma were reported in 1988, an incidence rate of 15.4 cases per 10,000 full-time workers. Ninety percent of these occurred in manufacturing.

Disorders associated with repeated trauma have become more frequently reported. In 1983, these disorders accounted for 25% of reported occupational illnesses in the private sector. They have significantly increased in frequency and as a percent of occupational illnesses each year and, in 1988, accounted for 48% of reported occupational illnesses (Occupational Illnesses and Injury, 1988).

Claims filed for the various categories included under cumulative trauma disorders from the Ohio Bureau of Workers' Compensation database for 1985-1989 are indicated in Table 4. In 1989, 6,485 claims were filed for cumulative trauma disorders accounting for 4.5% of all claims. Claims resulting from cumulative trauma disorders increased 84.6% between 1985 and 1989, an annual rate of increase of 16.6%. Among specified disorders over the five-year period, sizable increases were noted for carpal tunnel syndrome (127.7%) and disorders related to tendons, such as tenosynovitis and deQuervain's disease (40.6%). Carpal tunnel syndrome accounted for 49.1% of reported cumulative trauma disorder claims in Ohio in 1989.

A population-based study of the incidence of occupational carpal tunnel syndrome using Washington State Workers Compensation data found an overall incidence rate of 1.74 claims per 1,000 full-time equivalent employees (FTE) over the 1984 through 1988 period (Franklin, et al. 1991).

Table 4: The Number of Claims Filed for Inflammation, Irritation of Joints, Tendons or Muscles: Ohio, 1985-1989

	*Number of Claims**				
Code description	1985	1986	1987	1988	1989
Inflammation, irritation of the joints, tendons or muscles - unspecified	**551**	**488**	**649**	**626**	**780**
Disorders related to the joints (joint fatigue, osteoarthritis, rheumatoid arthritis, arthritis)	**49**	**64**	**45**	**51**	**53**
Disorders related to tendons, (tenosynovitis, de Quervain's disease, ganglions, tendonitis, synovitis)	**1,036**	**1,080**	**1,108**	**1,255**	**1,457**
Epicondylitis (tennis elbow)	**173**	**169**	**164**	**179**	**261**
Neuritis (thoracic outlet syndrome, inflammation of nerve endings)	**25**	**20**	**38**	**22**	**24**
Carpal tunnel syndrome	**1,399**	**1,688**	**1,882**	**2,373**	**3,186**
Bursitis	**140**	**106**	**100**	**91**	**96**
Vibration white finger (Raynaud's syndrome)	**7**	**6**	**4**	**3**	**6**
Inflammations, irritation of the joints, tendons or muscles - not elsewhere classified	**133**	**285**	**273**	**373**	**622**
Total	**3,513**	**3,906**	**4,263**	**4,973**	**6,485**

**Based on claims filed in the indicated calendar year under the Ohio Workers' Compensation law involving one or more days away from work and all occupational disease claims, whether or not they involved lost time.*

Source: Statistical Research Section, Division of Safety and Hygiene, Ohio Bureau of Workers' Compensation.

Annual incidence rates were 1.96 per 1,000 FTEs for women and 1.58 per 1,000 for men, a female to male incidence ratio of approximately 1.2:1. These data reflect all workers covered by workers' compensation in Washington and included approximately 1.3 million full-time workers in 1988.

Incidence rates increased from 1.78 per 1,000 FTEs in 1984 to 2.00 per 1,000 in 1988. Rates for women increased significantly over this period, while male incidence rates were relatively stable, with no statistically significant trend.

In addition to disability and work loss, occupationally related cumulative trauma disorders, such as carpal tunnel syndrome, affect the use of health care resources. The Centers for Disease Control, based on physician reporting, estimates that approximately 47% of carpal tunnel syndrome cases are work-related (MMWR, 1989, No. 38). Applying this percentage to national estimates indicates that each year more than 103,000 carpal tunnel release operations result from occupational disease.

References

Bureau of Labor Statistics, Occupational Injuries and Illnesses in the United States by Industry, 1988. Bulletin 2368. August 1990.

Bureau of Labor Statistics, Injury and Illness Data Available from 1987 Workers Compensation Records, Announcement 90-1, May 1990.

Centers for Disease Control, "Occupational Disease Surveillance: Carpal Tunnel Syndrome." MMWR, 1989, No. 38.

Franklin GM, Haug J, Heyer N, Checkoway H, Peck N, "Occupational Carpal Tunnel Syndrome in Washington State, 1984-1988," American Journal of Public Health, June 1991.

National Safety Council, Accident Facts, 1991 Edition. Chicago.

Ohio Bureau of Workers' Compensation, Occupational Injury and Illness Statistics, 1990. (Other data relating to back injuries and cumulative trauma disorders were provided by the Statistical Research Section, Division of Safety and Hygiene).

chapter 5

Impact

Apart from the extensive use of health care resources by those with musculoskeletal injuries their impact can also be measured by the extent to which they limit or restrict the normal activities of daily living.

Of the 32.8 million musculoskeletal injuries that were reported on average each year between 1985 and 1988, 29.4 million, or 89.7%, were medically attended. Slightly over half, or 53.8%, resulted in activity restrictions and 7.3 million, or 22.3%, resulted in bed-disability (Table 1). An activity-restricting injury is defined by the National Health Interview Survey (NHIS) as an injury that causes at least one-half day of restricted activity (reduction in a person's activity below his or her normal level). A bed-disability injury is defined as an injury that results in at least one half day of bed-disability.

Table 1: Average Annual Number of Injuries and Percent Medically Attended[1], Activity Restricting and Bed Disabling, by Type of Injury: United States 1985-88

	(thousands)				*(percent)*			
	Total	Medically Attended	Activity Restricting	Bed Disabling	Total	Medically Attended	Activity Restricting	Bed Disabling
All injuries	**61,137**	**54,683**	**30,573**	**13,352**	**100**	**89.4**	**50.0**	**21.8**
Musculoskeletal injuries	**32,780**	**29,406**	**17,639**	**7,315**	**100**	**89.7**	**53.8**	**22.3**
Fractures	**6,155**	**5,935**	**4,094**	**1,812**	**100**	**96.4**	**66.5**	**29.4**
Hip	***159**	***159**	***117**	***117**	***100**	***100.0**	***73.8**	***73.8**
Neck and trunk	**857**	**799**	**545**	**357**	**100**	**93.3**	**63.6**	**41.7**
Humerus, radius and ulna	**759**	**759**	**497**	**201**	**100**	**100.0**	**65.5**	**26.5**
Tibia, fibula and ankle	**492**	**443**	**452**	**279**	**100**	**90.1**	**91.8**	**56.8**
Other limb	**2,958**	**2,958**	**1,875**	**603**	**100**	**100.0**	**63.4**	**20.4**
Dislocations and sprains	**14,667**	**12,128**	**9,055**	**3,958**	**100**	**82.7**	**61.7**	**27.0**
Open wounds	**9,480**	**9,178**	**3,323**	**1,097**	**100**	**96.8**	**35.1**	**11.6**
Other injuries	**2,189**	**1,928**	**1,051**	**417**	**100**	**88.1**	**48.0**	**19.1**

[1]Medically attended injuries and activity restricting injuries are not mutually exclusive. Bed Disabling injuries are activity restricting.
**Estimate does not meet standards of reliability or precision.*
Note: Includes injuries which are either medically attended or result in at least one-half day of restricted activity.

Source: *National Center for Health Statistics, National Health Interview Surveys, Data tapes 1985-1988.*

Injury categories

Among the major categories of musculoskeletal injuries, fractures and open wounds were most likely to be medically attended (96.4% and 96.8%, respectively). Fractures were most likely to result in activity restriction or bed-disability. Almost two-thirds (66.5%) of fractures resulted in a person's restricting his or her normal activities and 29.4% resulted in bed-disability. Dislocations and sprains were the second leading cause of restrictions on activity, with 61.7% resulting in activity restrictions and 27.0% in bed-disability.

Activity limitation

Activity limitation associated with fractures varies with anatomic site. As would be expected because of restrictions on mobility, lower extremity fractures (excluding foot) were most likely to result in activity restrictions (91.8%) and bed-disability (56.8%).

Musculoskeletal injuries result in a substantial number of days of restricted activity and bed-disability, as well as work or school loss. Musculoskeletal injuries were responsible for over 249 million restricted activity days and over 71 million bed-disability days annually in the United States from 1985 through 1988. The percent distribution of restricted activity days and bed-disability resulting from musculoskeletal injuries is indicated in Figure 1.

Figure 1: Restricted Activity Days and Bed Disability Days Resulting from Musculoskeletal Injuries by Type of Injury

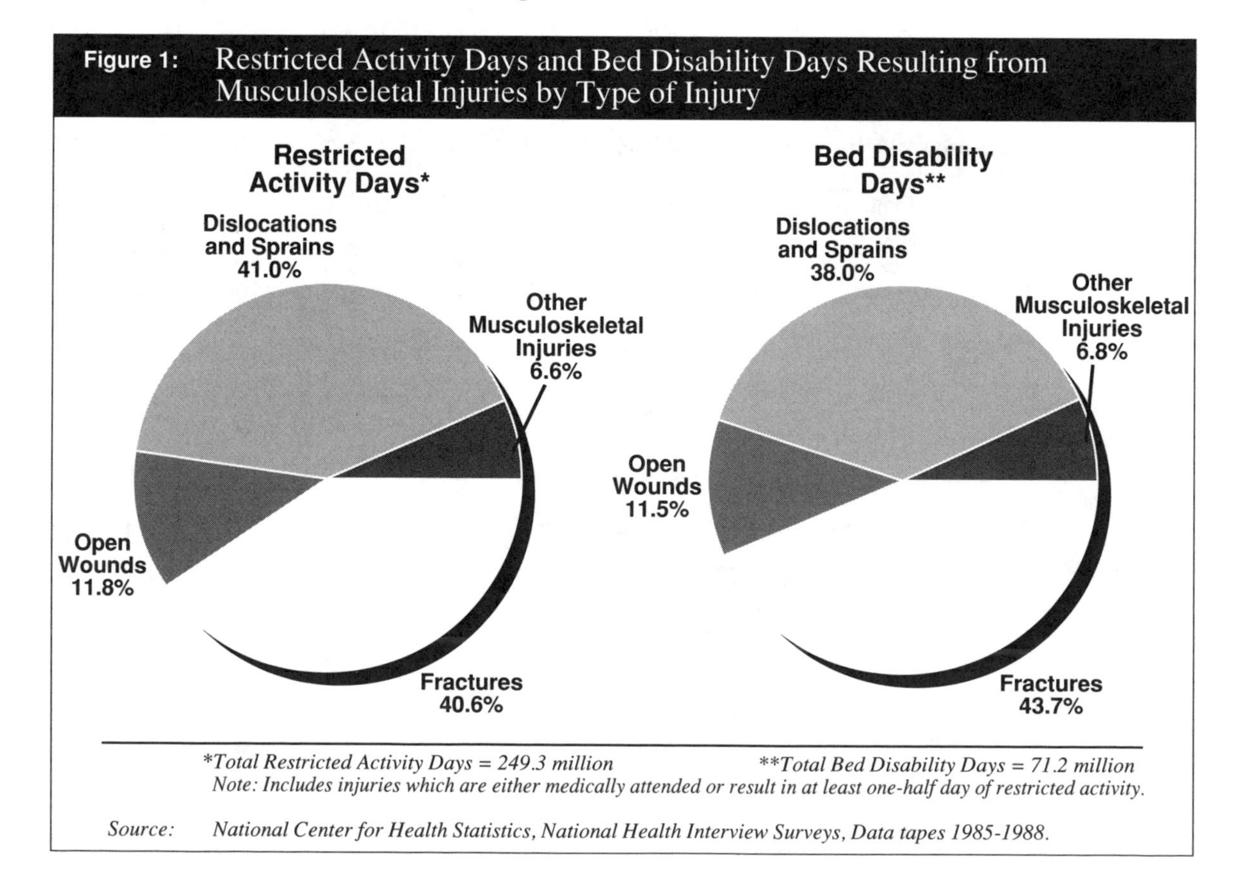

**Total Restricted Activity Days = 249.3 million* ***Total Bed Disability Days = 71.2 million*
Note: Includes injuries which are either medically attended or result in at least one-half day of restricted activity.

Source: National Center for Health Statistics, National Health Interview Surveys, Data tapes 1985-1988.

Dislocations and sprains were associated with the largest proportion of restricted activity days (41.0%), with fractures accounting for an additional 40.6% (Figure 1). Smaller percentages were attributable to open wounds (11.8%) and other musculoskeletal injuries (6.6%).

In relation to bed-disability days, fractures accounted for a plurality, 43.7%, followed by dislocations and sprains with 38.0%. Open wounds accounted for 11.5% of bed-disability days and other musculoskeletal injuries 6.8%.

The average duration of restricted activity associated with a musculoskeletal injury was 7.6 days (Table 2). Men reported 134.4 million restricted activity days and women reported 114.9 million restricted activity days. The restricted activity per injury was longer for women (8.2 days) than for men (7.2 days). Age was a major factor in determining the average duration of restricted activity for musculoskeletal injuries. For persons under 65 years of age, there were 210.7 million days of restricted activity for these injuries, with the duration of the restricted activity of 6.9 days. For persons 65 years of age and older, there were 38.6 million days of restricted activity; however, the average duration of this restricted activity period was 16.1 days.

Fractures resulted in 101 million days of restricted activity and the longest average duration of restricted activity per episode among all the injury categories (16.5 days). Men with fractures reported 52.0 million days of restricted activity, for an average duration of 15.5 days. The corresponding numbers for women were 49.3 million and 17.6 days. As was true for all musculoskeletal injuries, age was a major factor in determining the length of the duration of restricted activity associated with fractures. For persons under 65 years of age, the number of restricted activity days was 78 million, for an average duration of 2 weeks (14.0 days). For persons 65 years of age and older, restricted activity days totaled 23.4 million, with an average duration of 38.7 days, which is more than two-and-one-half times as long as those younger than 65 years of age.

Duration of restricted activities

The average duration of restricted activity was less for other types of musculoskeletal injuries, such as dislocations and sprains and open wounds. Dislocations and sprains resulted in an average per episode of 7.0 days of restricted activity and open wounds an average of 3.1 days of restricted activity per episode. There are only minor differences by gender. Those age 65 and older experience a longer duration of restricted activity, but the disparity was not proportionately as great as with fractures.

Bed disability

Musculoskeletal injuries also resulted in an average of 2.2 bed disability days per episode (Table 2). The general pattern of bed disability days is similar to that of restricted activity days. Although the distribution of bed disability days by gender was fairly equal, 36.0 million for men and 35.2 million for women, the average duration per injury episode was substantially longer for women (2.5 days) than for men (1.9 days). The average duration per episode was also significantly longer among those 65 and older (5.5 days), than those under 65 (1.9 days).

Among the various types of musculoskeletal injuries, fractures result in the longest average duration of bed days. For all persons, fractures resulted in 31.1 million bed-disability days, anaverage duration per episode of 5.0 days. Bed days per episode did not vary substantially between men (4.9 days) and women (5.2 days). There was, however, a substantial variation by age category. The average durations of bed disability per fracture was 15.4 days for those 65 and older compared with 3.9 days for those under 65. Although only 9.8% of fractures occur in the 65 and older population, because of the disparity in average duration, 30.0% of bed disability days associated with fractures occur in the 65 and older age group.

Work-loss

Musculoskeletal injuries also result in a substantial number of work-loss and school-loss days. Musculoskeletal injuries were responsible for over 77.6 million work loss days and 7.3 million school loss days annually between 1985 and 1988. The percent distributions of work-loss and school-loss days resulting from the different types of musculoskeletal injuries are indicated in Figure 2.

Dislocations and sprains were responsible for almost half (46.9%) of work-loss days associated with musculoskeletal injuries. Fractures were responsible for 33.6% of work-loss days and open wounds for 12.9%.

Among the different types of musculoskeletal injuries, fractures were responsible for the largest number (40.3%) of school loss days. Dislocations and sprains were responsible for 36.4% of school loss days and open wounds for 16.8%.

Table 2: Restricted Activity Days and Bed disability Days Associated with an Injury by Gender and Age: United States, 1985-1988

	Average Annual Incidence (thousands)	Restricted Activity Days (thousands)	Days/ Injury	Bed Disability Days (thousands)	Days/ Injury
Total					
All injuries[1]	61,137	414,888	6.8	133,239	2.2
Musculoskeletal injuries[2]	32,780	249,329	7.6	71,154	2.2
Fractures[3]	6,155	101,261	16.5	31,068	5.0
Sprains[4]	14,667	102,141	7.0	27,011	1.8
Open wounds[5]	9,480	28,118	3.0	7,752	0.8
Crushing injury[6]	289	1,276	4.4	490	1.7
Other injuries[7]	2,189	16,533	7.6	4,833	2.2
Males					
All injuries[1]	33,466	208,849	6.2	62,043	1.9
Musculoskeletal injuries[2]	18,766	134,429	7.2	35,986	1.9
Fractures[3]	3,361	51,995	15.5	16,534	4.9
Sprains[4]	7,860	55,286	7.0	13,200	1.7
Open wounds[5]	6,181	17,484	2.8	3,626	0.6
Crushing injury[6]	*192	1,016	*5.3	323	*1.7
Other injuries[7]	1,171	8,648	7.4	2,303	2.0
Females					
All injuries[1]	27,671	206,039	7.4	71,196	2.6
Musculoskeletal injuries[2]	14,014	114,900	8.2	35,168	2.5
Fractures[3]	2,794	49,266	17.6	14,534	5.2
Sprains[4]	6,807	48,855	6.9	13,810	2.0
Open wounds[5]	3,299	10,634	3.2	4,126	1.3
Crushing injury[6]	*96	260	*2.7	167	*1.7
Other injuries[7]	1,018	7,885	7.7	2,530	2.5
Less than age 65					
All injuries[1]	55,577	334,669	6.0	102,776	1.8
Musculoskeletal injuries[2]	30,388	210,713	6.9	57,976	1.9
Fractures[3]	5,551	77,891	14.0	21,737	3.9
Sprains[4]	13,739	91,968	6.7	24,796	1.8
Open wounds[5]	8,879	25,220	2.8	6,383	0.7
Crushing injury[6]	273	1,116	*4.1	410	1.5
Other injuries[7]	1,945	14,519	7.5	4,650	2.4
65 years and over					
All injuries[1]	5,561	80,220	14.4	30,464	5.5
Musculoskeletal injuries[2]	2,393	38,615	16.1	13,178	5.5
Fractures[3]	604	23,371	38.7	9,331	15.4
Sprains[4]	928	10,172	11.0	2,215	2.4
Open wounds[5]	601	2,898	4.8	1,370	2.3
Crushing injury[6]	*15	160	*10.7	80	*5.3
Other injuries[7]	244	2,014	8.3	183	0.8

**Estimate does not meet standards of reliability or precision.*
Note: Includes injuries which are either medically attended or result in at least one-half day of restricted activity.
[1]ICD-9-CM 800-999
[2]ICD-9-CM 805-829,831-848, 926-928, 874-877, 879-884, 890-894, 954-957, 959
[3]ICD-9-CM 805-829
[4]ICD-9-CM 831-848
[5]ICD-9-CM 846-847
[6]ICD-9-CM 874-877, 879-884, 890-894
[7]ICD-9-CM 954-957, 959
Source: *National Center for Health Statistics, National Health Interview Surveys, Data tapes 1985-1988.*

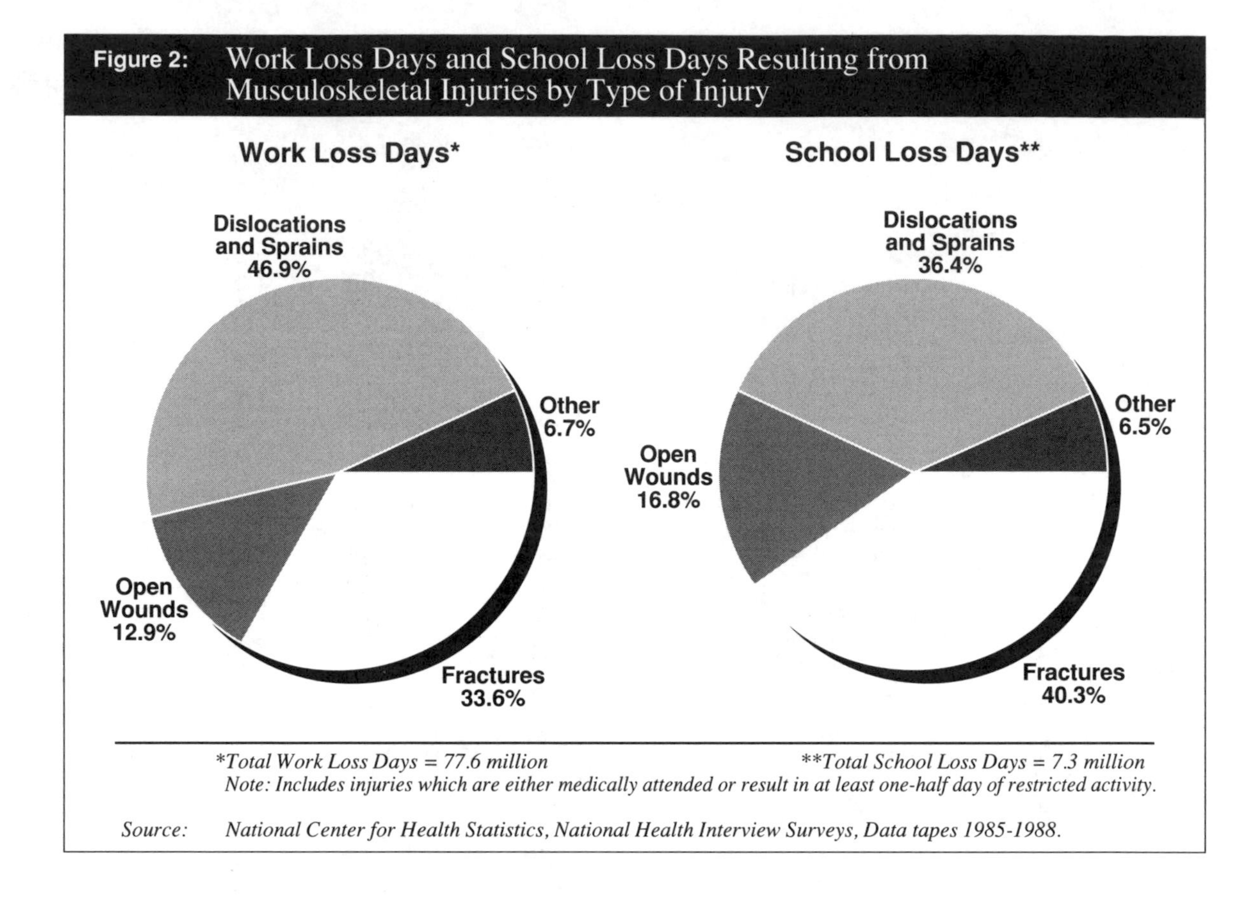

Figure 2: Work Loss Days and School Loss Days Resulting from Musculoskeletal Injuries by Type of Injury

**Total Work Loss Days = 77.6 million* ***Total School Loss Days = 7.3 million*

Note: Includes injuries which are either medically attended or result in at least one-half day of restricted activity.

Source: National Center for Health Statistics, National Health Interview Surveys, Data tapes 1985-1988.

Reflecting to a large extent the greater labor force participation rate of men than women, 65.5% of musculoskeletal injuries resulting in work loss days occurred among men (Table 3). The average duration of work-loss days associated with musculoskeletal injuries was 4.6 days. Men experienced a slightly longer average work-loss (4.7 days) than women (4.3 days). The difference was attributable primarily to a longer average duration of work-loss resulting from sprains (4.7 vs. 4.2) and open wounds (2.1 vs. 1.5) among men than women).

School-loss

School-loss days, as shown in Table 4, are an important measure of the impact of injuries on the school-age population 5 to 17 years of age. Musculoskeletal injuries resulted in an average duration of school-loss of approximately 1 day (0.9 days). The average duration was longer among boys (1.0 days) than girls (0.7 days). The largest variation in average duration by gender occurred in relation to fractures for which the average duration among boys was 1.8 days compared with 1.2 days among girls.

Table 3: Work Loss Days Associated with an Injury for Persons Currently Employed, by Gender, United States: 1985-1988

	Average Annual Incidence (thousands)	Work Loss Days (thousands)	Days/ Injury
Total			
All injuries[1]	28,747	114,595	4.0
Musculoskeletal injuries[2]	17,049	77,637	4.6
Fractures[3]	2,548	26,052	10.2
Sprains[4]	8,098	36,405	4.5
Open wounds[5]	5,020	9,318	1.9
Crushing injury[6]	209	664	3.2
Other injuries[7]	1,174	5,198	4.4
Males			
All injuries[1]	17,550	72,565	4.1
Musculoskeletal injuries[2]	10,868	50,816	4.7
Fractures[3]	1,718	17,543	10.2
Sprains[4]	4,740	22,336	4.7
Open wounds[5]	3,506	7,216	2.1
Crushing injury[6]	153	478	3.1
Other injuries[7]	751	3,244	4.3
Females			
All injuries[1]	11,197	42,030	3.8
Musculoskeletal injuries[2]	6,181	26,821	4.3
Fractures[3]	830	8,509	10.3
Sprains[4]	3,358	14,069	4.2
Open wounds[5]	1,514	2,102	1.4
Crushing injury[6]	*56	186	*3.3
Other injuries[7]	423	1,954	4.6

**Estimate does not meet standards of reliability or precision.*
Note: Includes injuries which are either medically attended or result in at least one-half day of restricted activity.

[1] *ICD-9-CM 800-999*
[2] *ICD-9-CM 805-829,831-848, 926-928, 874-877, 879-884, 890-894, 954-957, 959*
[3] *ICD-9-CM 805-829*
[4] *ICD-9-CM 831-848*
[5] *ICD-9-CM 846-847*
[6] *ICD-9-CM 874-877, 879-884, 890-894*
[7] *ICD-9-CM 954-957, 959*

Source: *National Center for Health Statistics, National Health Interview Surveys, Data tapes 1985-1988.*

Table 4: School Loss Days Associated with an Injury for Persons Currently Employed, by Gender, United States: 1985-1988

	Average Annual Incidence (thousands)	School Loss Days (thousands)	Days/ Injury
Total			
All injuries[1]	14,509	12,848	0.9
Musculoskeletal injuries[2]	8,064	7,377	0.9
Fractures[3]	1,845	2,958	1.6
Sprains[4]	3,299	2,667	0.8
Open wounds[5]	2,515	1,079	0.4
Crushing injury[6]	*42	154	*3.7
Other injuries[7]	362	479	1.3
Males			
All injuries[1]	8,524	8,022	0.9
Musculoskeletal injuries[2]	4,908	5,010	1.0
Fractures[3]	1,177	2,127	1.8
Sprains[4]	1,850	1,617	0.9
Open wounds[5]	1,594	787	0.5
Crushing injury[6]	*15	154	*10.3
Other injuries[7]	272	326	1.2
Females			
All injuries[1]	5,986	4,826	0.8
Musculoskeletal injuries[2]	3,156	2,327	0.7
Fractures[3]	668	831	1.2
Sprains[4]	1,450	1,050	0.7
Open wounds[5]	1,594	787	0.5
Crushing injury[6]	*15	154	*10.3
Other injuries[7]	90	153	1.7

**Estimate does not meet standards of reliability or precision.*
Note: Includes injuries which are either medically attended or result in at least one-half day of restricted activity.

[1]ICD-9-CM 800-999
[2]ICD-9-CM 805-829,831-848, 926-928, 874-877, 879-884, 890-894, 954-957, 959
[3]ICD-9-CM 805-829
[4]ICD-9-CM 831-848
[5]ICD-9-CM 846-847
[6]ICD-9-CM 874-877, 879-884, 890-894
[7]ICD-9-CM 954-957, 959

Source: *National Center for Health Statistics, National Health Interview Surveys, Data tapes 1985-1988.*

References

National Center for Health Statistics, National Health Interview Survey, 1985-1988 (Data Tapes).

chapter 6

Health Care Utilization

Musculoskeletal injuries result in a substantial use of health care resources and are responsible for large numbers of hospitalizations and physician visits each year. In relation to inpatient care, during the period from 1985 through 1988, more than 1.5 million hospitalizations were recorded annually in the United States for all types of musculoskeletal injuries (Table 1) and accounted for 45.1% of hospitalizations for musculoskeletal conditions. These injuries resulted in almost 11 million patient days per year, with an average length of stay of 7.1 days.

Table 1: Average Annual Hospitalizations Resulting from Musculoskeletal Injuries, United States: 1985-1988 by Age and Gender[1]

	Gender	Total	Less than 18	18-44	45-64	65 & over	Hospital Days[2]	Average Length of Stay
Fractures		**930,000**	**127,000**	**261,000**	**149,000**	**394,000**	**8,305,000**	**8.9**
	Male	**434,000**	**89,000**	**181,000**	**71,000**	**94,000**	**3,393,000**	**7.8**
	Female	**496,000**	**38,000**	**79,000**	**78,000**	**300,000**	**4,912,000**	**9.9**
Dislocations		**80,000**	**10,000**	**45,000**	**17,000**	**9,000**	**283,000**	**3.5**
	Male	**52,000**	**6,000**	**33,000**	**9,000**	**3,000**	**159,000**	**3.1**
	Female	**28,000**	**4,000**	**12,000**	**7,000**	**6,000**	**124,000**	**4.4**
Sprains and strains		**238,000**	**15,000**	**132,000**	**62,000**	**29,000**	**1,044,000**	**4.4**
	Male	**131,000**	**8,000**	**81,000**	**31,000**	**12,000**	**536,000**	**4.1**
	Female	**106,000**	**7,000**	**52,000**	**31,000**	**17,000**	**508,000**	**4.8**
Crushing injuries		**10,000**	*	**6,000**	*	*	**72,000**	**7.3**
	Male	**9,000**	*	**6,000**	*	*	**63,000**	**7.1**
	Female	*	*	*	*	*	**9,000**	**9.9**
Contusions		**77,000**	**12,000**	**32,000**	**12,000**	**20,000**	**323,000**	**4.2**
	Male	**39,000**	**7,000**	**21,000**	**6,000**	**5,000**	**135,000**	**3.5**
	Female	**38,000**	**5,000**	**11,000**	**7,000**	**15,000**	**188,000**	**5.0**
Open wounds		**157,000**	**27,000**	**102,000**	**18,000**	**10,000**	**677,000**	**4.3**
	Male	**123,000**	**20,000**	**84,000**	**14,000**	**5,000**	**516,000**	**4.2**
	Female	**35,000**	**7,000**	**18,000**	**4,000**	**6,000**	**161,000**	**4.6**
Other injuries		**50,000**	**8,000**	**28,000**	**8,000**	**6,000**	**196,000**	**4.0**
	Male	**34,000**	**5,000**	**21,000**	**6,000**	**3,000**	**144,000**	**4.2**
	Female	**15,000**	**3,000**	**7,000**	*	**3,000**	**52,000**	**3.4**
Total, all musculoskeletal injuries		**1,541,000**	**200,000**	**606,000**	**267,000**	**468,000**	**10,900,000**	**7.1**
	Male	**822,000**	**136,000**	**426,000**	**138,000**	**121,000**	**4,947,000**	**6.0**
	Female	**720,000**	**64,000**	**179,000**	**129,000**	**347,000**	**5,954,000**	**8.3**

[1]*First listed diagnosis for inpatients discharged from short-stay hospitals*
[2]*Annual Average*
**Estimate does not meet standards of reliability or precision.*

Source: *National Center for Health Statistics, National Hospital Discharge Survey, Data tapes, 1985-1988.*

The number of hospitalizations was higher for males in all injury categories except fractures. Lengths of stay, however, were higher for females overall and in all categories except other injuries.

Injuries requiring hospitalization

Comparing the aggregate categories, fractures were the dominant factor in use of inpatient services. Fractures accounted for 60.4% (Table 2) of hospitalizations for musculoskeletal injuries in short-stay hospitals with, on average, 930,000 occurring each year. Patient days were even more concentrated with fractures accounting for 76.2% of patient days associated with musculoskeletal injuries. Among the musculoskeletal injury categories, fractures also required the longest stay at 8.9 days.

Table 2: Distribution of Hospitalizations & Patient Days Resulting from Musculoskeletal Injuries: United States, 1985-1988

	Percent	
	Hospitalizations[1]	Patient days[2]
Fractures	**60.4**	**76.2**
Dislocations	**5.2**	**2.6**
Sprains and strains	**15.4**	**9.6**
Crushing injury	***0.6**	***0.7**
Contusions	**5.0**	**3.0**
Open wound	**10.2**	**6.2**
Other injury	**3.2**	**1.8**
Total	**100.0**	**100.0**

[1]*First listed diagnosis for inpatients discharged from short-stay hospitals*
[2]*Annual Average*
**Estimate does not meet standards of reliability or precision.*

Source: *National Center for Health Statistics, National Hospital Discharge Survey, Data tapes, 1985-1988.*

Fractures resulted in especially heavy use of inpatient services among women. As indicated above, fractures were the only major injury category where the majority of hospitalizations (53.3%) occurred among women. Combined with a longer length of stay, 9.9 days for women compared with 7.8 days for men, patient days resulting from fractures accounted for 82.5% of the patient days associated with musculoskeletal injuries among women compared with 68.6% among men.

Sprains and strains accounted for the second largest number of hospitalizations (238,000) for musculoskeletal injuries. Dislocations resulted in comparatively few hospitalizations: only 80,000 annually. Lengths of stay for sprains (4.4 days) and dislocations (3.5 days) were less than half that of fractures.

The remaining categories: crushing injuries, contusions, open wounds, and other injuries accounted for less than 20% of hospitalizations for musculoskeletal injuries. Most of these were the result of open wounds (157,000) or contusions (77,000).

Injuries by anatomic site

Hospitalizations resulting from fractures and from sprains or dislocations at selected anatomic sites are indicated in Tables 3 and 4. Fractures leading to hospitalizations were most likely to involve the neck of the femur, radius and ulna, and vertebral column. The number of hospitalizations for fractures among women were disproportionately large for fractures occurring at three sites: fractures of the neck of the femur, the pelvis and the humerus. The greater frequency occurs primarily among those 65 and older and reflects not only the larger female population in this age group but, also, the susceptibility to fracture of these bones among the older female population as a result of osteoporosis. Approximately 86.1% of fractures of the neck of the femur, for instance, occur in the 65 and older group and 76.7% of these cases are among women.

The average length of stay was generally shorter for fractures occurring at upper extremity sites than for those of either the trunk or lower extremities. Average lengths of stay were longest for femur fractures other than the neck of the femur (16.2 days), fractures of the neck of the femur (13.7 days), and fractures of the pelvis (12.2 days).

Hospitalizations associated with sprains primarily involve the back (101,000) and knee (40,000). Back sprains alone account for 42.6% of hospitalizations associated with sprains and 53.2% of patient days (Table 4). The average length of stay for hospitalizations as a result of back sprains was 5.5 days.

For dislocations, hospitalizations primarily involve dislocations of the knee (44,000) or shoulder (12,000). Lengths of stay, at 2.6 days, were relatively short for each of these conditions.

Injuries requiring physician visits by category

Musculoskeletal injuries also result in large numbers of physician visits. In 1985, more than 37.2 million visits to physicians in office-based practice resulted from musculoskeletal injuries (Table 5).

Table 3: Average Annual Hospitalizations Resulting from Fractures at Selected Anatomic Sites: United States, 1985-1988, by Age and Gender[1]

	Gender	Total	Less than 18	18-44	45-64	65 & over	Hospital Days[2]	Average Length of Stay
Fracture of hand		30,000	4,000	19,000	5,000	*	80,000	2.7
	Male	23,000	3,000	15,000	3,000	*	59,000	2.5
	Female	6,000	*	4,000	*	*	21,000	3.3
Fracture of foot		30,000	3,000	16,000	7,000	3,000	155,000	5.2
	Male	22,000	3,000	13,000	5,000	*	114,000	5.1
	Female	8,000	*	3,000	2,000	2,000	42,000	5.3
Fracture of ribs and sternum		49,000	*	14,000	13,000	20,000	280,000	5.7
	Male	26,000	*	10,000	8,000	7,000	143,000	5.5
	Female	23,000	*	5,000	5,000	13,000	137,000	6.0
Fracture of carpals		8,000	*	5,000	*	*	30,000	3.9
	Male	6,000	*	4,000	*	*	22,000	3.9
	Female	2,000	*	*	*	*	8,000	4.0
Fracture of Clavicle or scapula		12,000	*	7,000	*	*	54,000	4.5
	Male	7,000	*	4,000	*	*	28,000	4.1
	Female	5,000	*	3,000	*	*	26,000	5.1
Fracture of radius and ulna		88,000	25,000	29,000	18,000	17,000	334,000	3.8
	Male	48,000	17,000	21,000	7,000	4,000	182,000	3.8
	Female	41,000	8,000	9,000	11,000	13,000	152,000	3.7
Fracture of humerus		66,000	24,000	11,000	9,000	22,000	363,000	5.5
	Male	29,000	14,000	8,000	4,000	3,000	147,000	5.0
	Female	37,000	10,000	3,000	5,000	19,000	216,000	5.9
Fracture of tibia and fibula		77,000	17,000	34,000	12,000	14,000	569,000	7.4
	Male	49,000	14,000	26,000	7,000	3,000	341,000	7.0
	Female	28,000	3,000	8,000	5,000	11,000	229,000	8.2
Fracture of vertebral column		79,000	6,000	30,000	14,000	29,000	740,000	9.4
	Male	42,000	4,000	21,000	9,000	9,000	425,000	10.1
	Female	37,000	2,000	10,000	5,000	20,000	316,000	8.6
Fracture of neck of femur		252,000	4,000	8,000	23,000	217,000	3,462,000	13.7
	Male	67,000	2,000	6,000	8,000	51,000	896,000	13.3
	Female	185,000	*	2,000	14,000	167,000	2,556,000	13.9
Other fracture of femur		58,000	21,000	16,000	7,000	15,000	938,000	16.2
	Male	33,000	16,000	11,000	3,000	3,000	484,000	14.8
	Female	25,000	5,000	5,000	3,000	12,000	454,000	18.1
Fracture of pelvis		51,000	4,000	14,000	6,000	28,000	618,000	12.2
	Male	19,000	2,000	7,000	3,000	7,000	254,000	13.3
	Female	32,000	*	6,000	3,000	21,000	363,000	11.5

[1]First listed diagnosis for inpatients discharged from short-stay hospitals.
[2]Annual Average
**Estimate does not meet standards of reliability or precision.*

Source: National Center for Health Statistics, National Hospital Discharge Survey, Data tapes, 1985-1988.

Table 4: Average Annual Hospitalizations Resulting from Sprains or Dislocations at Selected Musculoskeletal Sites: United States, 1985-88, by Age and Gender[1]

	Gender	Total	Less than 18	18-44	45-64	65 & over	Hospital Days[2]	Average Length of Stay
Sprains:								
Sprains of wrist and hand		7,000	*	5,000	*	*	15,000	2.1
	Male	5,000	*	4,000	*	*	11,000	2.4
	Female	2,000	*	*	*	*	3,000	1.5
Sprains of ankle		11,000	*	7,000	2,000	*	29,000	2.8
	Male	8,000	*	6,000	*	*	19,000	2.4
	Female	2,000	*	*	*	*	10,000	4.0
Sprains of knee and leg		40,000	7,000	25,000	6,000	3,000	139,000	3.4
	Male	26,000	4,000	18,000	3,000	*	89,000	3.5
	Female	15,000	3,000	7,000	3,000	*	51,000	3.4
Sprains of back		101,000	*	59,000	28,000	13,000	556,000	5.5
	Male	51,000	*	35,000	12,000	4,000	267,000	5.2
	Female	50,000	*	24,000	16,000	9,000	289,000	5.8
Dislocations:								
Dislocations of shoulder		12,000	*	7,000	2,000	2,000	32,000	2.6
	Male	9,000	*	6,000	*	*	20,000	2.2
	Female	3,000	*	*	*	*	12,000	4.3
Dislocations of knee		44,000	5,000	26,000	10,000	3,000	113,000	2.6
	Male	27,000	2,000	18,000	5,000	*	66,000	2.5
	Female	17,000	3,000	8,000	5,000	*	47,000	2.8

[1]*First listed diagnosis for inpatients discharged from short-stay hospitals.*
[2]*Annual Average*
Estimate does not meet standards of reliability or precision.

Source: *National Center for Health Statistics, National Hospital Discharge Survey, Data tapes, 1985-1988.*

In contrast to injuries requiring inpatient care, sprains and strains were the most frequently occurring musculoskeletal injury and resulted in approximately 14.5 million physician visits. Fractures were the second most frequent category, resulting in 10.1 million visits, followed by open wounds (5,205,000), contusions (3,617,000) and dislocations (1,830,000). The relative distribution of office visits by type of musculoskeletal injury is indicated in Table 6.

Visits related to fractures were most frequent for fractures of the radius and ulna (1,816,000), hand (1,648,000) and foot (1,027,000) (Table 7). In contrast to visits for fractures, those relating to sprains and dislocations were more concentrated by anatomic site (Table 8). In relation to office

Table 5: Visits to Physicians in Office-based Practice for Musculoskeletal Conditions: United States, 1985, by Age and Gender[1]

(thousands)

	Gender	Total	Less than 18	18-44	45-64	65 & over
Fractures		**10,128**	**3,024**	**3,352**	**1,946**	**1,806**
	Male	**5,080**	**1,969**	**2,044**	**693**	**374**
	Female	**5,049**	**1,055**	**1,309**	**1,253**	**1,432**
Dislocations		**1,830**	**406**	**954**	**289**	*
	Male	**1,162**	*	**694**	*	*
	Female	**669**	*	**261**	*	*
Sprains and strains		**14,564**	**1,473**	**8,755**	**3,228**	**1,108**
	Male	**7,413**	**800**	**4,617**	**1,553**	**443**
	Female	**7,151**	**673**	**4,138**	**1,675**	**665**
Crushing injuries		***110**	*	*	*	*
	Male	*	*	*	*	*
	Female	*	*	*	*	*
Contusions		**3,617**	**747**	**1,523**	**729**	**618**
	Male	**1,521**	**316**	**712**	**287**	*
	Female	**2,096**	**431**	**811**	**442**	**412**
Open wounds		**5,205**	**1,540**	**2,230**	**1,047**	**387**
	Male	**3,256**	**999**	**1,499**	**571**	**187**
	Female	**1,949**	**541**	**731**	**476**	**200**
Other injuries		**1,762**	**322**	**972**	**351**	*
	Male	**1,248**	**191**	**738**	**257**	*
	Female	**514**	**130**	**234**	*	*
Total, all musculoskeletal injuries		**37,216**	**7,544**	**17,850**	**7,606**	**4,216**
	Male	**19,770**	**4,485**	**10,367**	**3,526**	**1,392**
	Female	**17,446**	**3,059**	**7,483**	**4,080**	**2,824**

[1]Principal diagnosis associated with patient's reason for visit.
**Estimate does not meet standards of reliability or precision.*

Source: *National Ambulatory Medical Care Survey, 1985.*

Table 6: Distribution of Visits to Physicians in Office-Based Practice for Musculoskeletal Injuries: United States, 1985[1]

	Percent
Fractures	**27.2**
Dislocations	**4.9**
Sprains and strains	**39.1**
Crushing injury	***0.3**
Contusions	**9.7**
Open wound	**14.0**
Other injuries	**4.7**
Total	**100.0**

[1]Principal diagnosis associated with patient's reason for visit.
**Estimate does not meet standards of reliability or precision.*

Source: *National Ambulatory Medical Care Survey, 1985.*

Table 7: Visits to Physicians in Office-Based Practice for Fractures at Selected Anatomic Sites: United States, 1985, by Gender[1]

	(thousands)		
	Total	Male	Female
Fracture of hand	1,648	1,181	467
Fracture of foot	1,027	462	566
Fracture of ribs and sternum	306	*	*
Fracture of carpals	607	328	279
Fracture of Clavicle or scapula	602	425	*
Fracture of radius and ulna	1,817	727	1,090
Fracture of humerus	621	285	336
Fracture of tibia and fibula	825	423	401
Fracture of vertebral column	352	*	*
Fracture of neck of femur	393	*	306
Other fracture of femur	278	*	*

[1]Principal diagnosis associated with patient's reason for visit.
**Estimate does not meet standards of reliability or precision.*

Source: *National Ambulatory Medical Care Survey, 1985.*

Table 8: Visits to Physicians in Office-Based Practice for Sprains or Dislocations at Selected Musculoskeletal Sites: United States, 1985, by Gender[1]

	(thousands)		
	Total	Male	Female
Sprains:			
Sprains of wrist and hand	717	327	390
Sprains of ankle	1601	726	875
Sprains of knee and leg	1,298	751	547
Sprains of back	5,592	3,147	2,445
Dislocations:			
Dislocations of shoulder	283	*	*
Dislocations of knee	1,252	796	457

[1]Principal diagnosis associated with patient's reason for visit.
**Estimate does not meet standards of reliability or precision.*

Source: *National Ambulatory Medical Care Survey, 1985.*

visits resulting from sprains, for example, back sprains were the most frequently indicated (5,592,000) and accounted for 38.4% of visits for sprains. For dislocations, 83.8% of visits resulted from dislocations of either the knee (1,252,000) or shoulder (283,000).

New problem office visits

Approximately 17.8 million new-problem office visits resulted from musculoskeletal injuries. The large majority of these (81.0%) were not referred to another physician (Table 9). New problem office visits without referral include all visits for new problems in which the patient was not referred to the index physician by another physician. New problem office visits with referral include only those visits for new problems in which the patient was referred to the index physician by another physician. Fractures were almost twice as likely to be referred (31.9%) as were the other categories of musculoskeletal injuries (16.2%).

Table 9: New Problem Office Visits Musculoskeletal Injuries: United States, 1985

	Without referral		*With referral*	
	Number (thousands)	Percent	Number (thousands)	Percent
All Musculoskeletal injuries	**14,410**	**81.0**	**3,378**	**19.0**
Fractures	**2,145**	**68.1**	**1,004**	**31.9**
Dislocations and sprains	**6,749**	**83.8**	**1,309**	**16.2**
All other musculoskeletal injuries*	**5,516**	**83.8**	**1,065**	**16.2**

**Crushing injury, contusion, open wound, other injuries.*

Source: *National Ambulatory Medical Care Survey, 1985.*

For the majority of cases, the initial contact physician for musculoskeletal injuries was a primary care physician (Figure 1). Of the 14.4 million new-problem office visits without referral*, 70.4 percent of visits were to general/family practitioners, internists or pediatricians. Approximately 17.6% were made to orthopaedic surgeons; 12.0% were made to other specialists.

For new-problem office visits for musculoskeletal injuries where a referral was made*, however, the majority of visits (59.8%) were made to

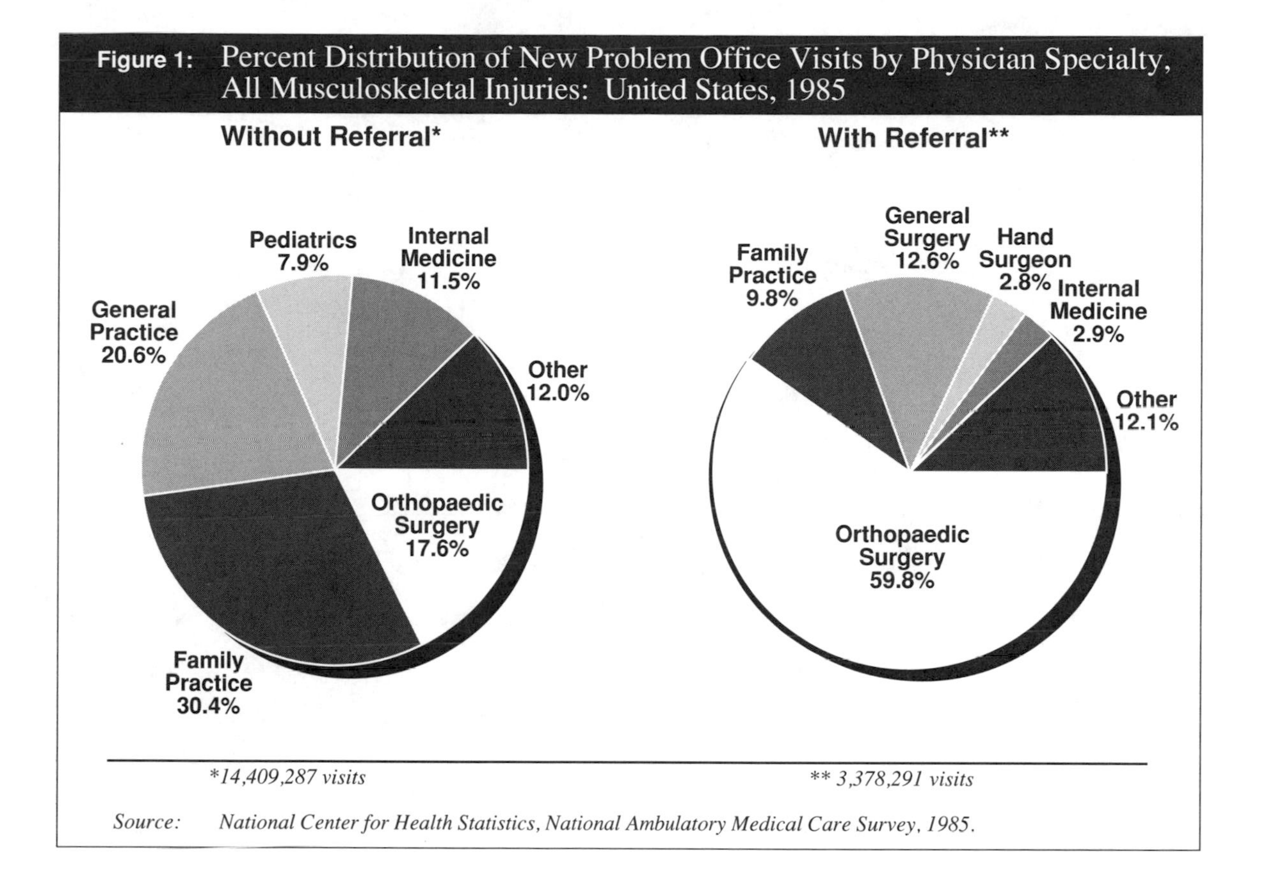

Figure 1: Percent Distribution of New Problem Office Visits by Physician Specialty, All Musculoskeletal Injuries: United States, 1985

Source: National Center for Health Statistics, National Ambulatory Medical Care Survey, 1985.

orthopaedic surgeons; an additional 12.6% to general surgeons. Less than 15% were referred to primary care physicians. As indicated in Figures 2-4, this pattern was consistent for the individual injury categories: fractures, dislocations and sprains, and back sprains. In relation to new-problem office visits without referral, for fractures, the majority of visits, 53.7%, were to primary care physicians (Figure 2); with referral 89.0% are made to orthopaedic surgeons.

For dislocations and sprains, 68.8% of visits without referral are made to primary care physicians; 20.0% to orthopaedic surgeons (Figure 3). Where a referral was made, 70.0% of new patient visits were made to orthopaedic surgeons. In relations to back sprains, 69.5% of initial visits were made to primary care physicians; of patients referred, 64.3% were referred to orthopaedic surgeons (Figure 4).

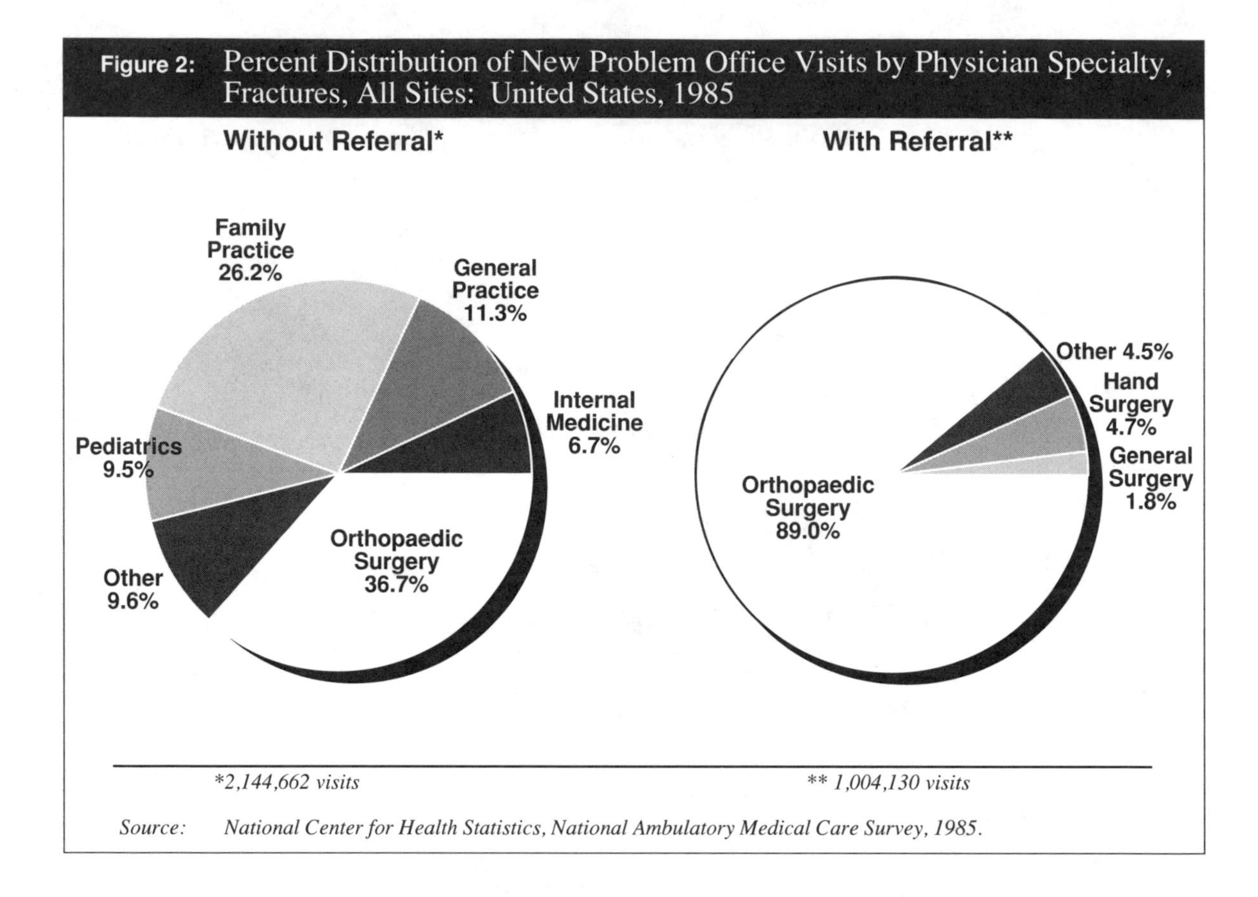

Figure 2: Percent Distribution of New Problem Office Visits by Physician Specialty, Fractures, All Sites: United States, 1985

**2,144,662 visits*

*** 1,004,130 visits*

Source: *National Center for Health Statistics, National Ambulatory Medical Care Survey, 1985.*

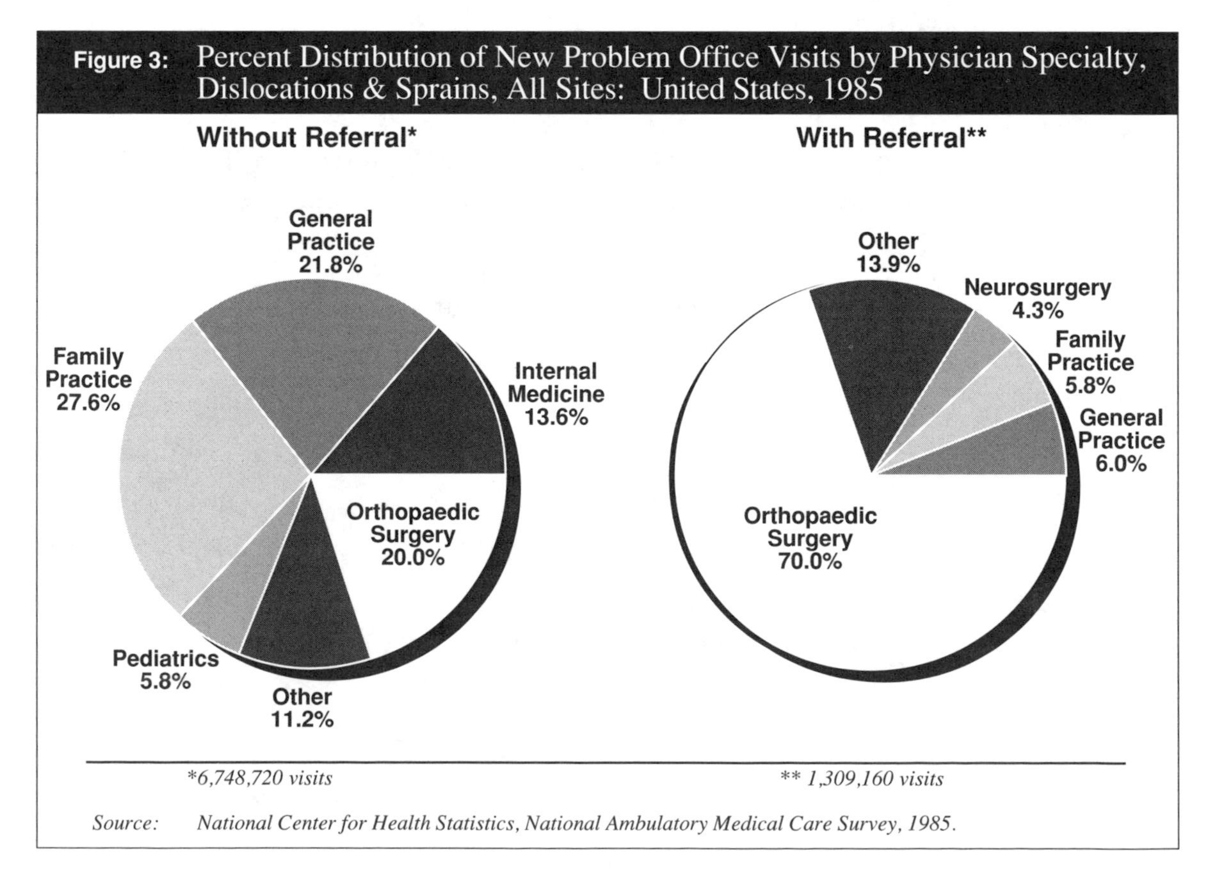

Figure 3: Percent Distribution of New Problem Office Visits by Physician Specialty, Dislocations & Sprains, All Sites: United States, 1985

**6,748,720 visits*

*** 1,309,160 visits*

Source: *National Center for Health Statistics, National Ambulatory Medical Care Survey, 1985.*

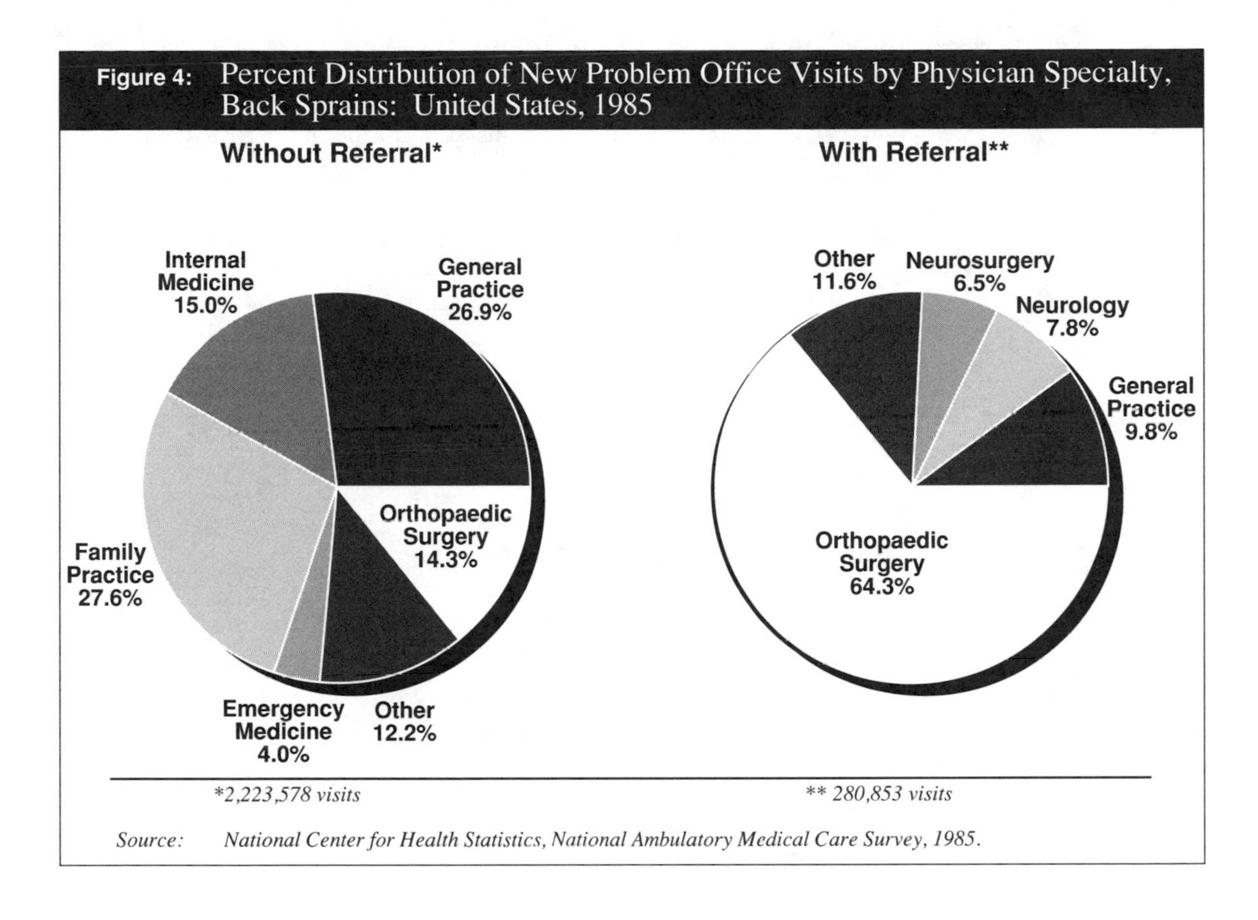

Figure 4: Percent Distribution of New Problem Office Visits by Physician Specialty, Back Sprains: United States, 1985
Without Referral*
Internal Medicine 15.0%
General Practice 26.9%
Family Practice 27.6%
Orthopaedic Surgery 14.3%
Emergency Medicine 4.0%
Other 12.2%
With Referral**
Other 11.6%
Neurosurgery 6.5%
Neurology 7.8%
General Practice 9.8%
Orthopaedic Surgery 64.3%
*2,223,578 visits
** 280,853 visits
Source: National Center for Health Statistics, National Ambulatory Medical Care Survey, 1985.

References

National Center for Health Statistics, National Ambulatory Medical Care Survey, 1985 (Data Tape).

National Center for Health Statistics, National Hospital Discharge Survey, 1985-1988 (Data Tape).

Section 4

Medical Implants and Major Joint Procedures

Medical Implants

An estimated 11 million persons in the United States report having at least one medical device implant (Table 1). Two types of implants, fixation devices and artificial joints, are used in the treatment of musculoskeletal diseases and injuries and account for 51.3% of all reported implants.

Table 1: Persons with One or More Medical Device Implants for Selected Types of Implants, by Age and Gender: United States, 1988

	Number of persons (thousands)	Percent of persons
All implants[1]		
All ages	**11,051**	**4.6**
65 years and older	**4,434**	**15.5**
85 years and older	**539**	**24.9**
Fixation device[2]		
All ages	**4,382**	**1.8**
65 years and older	**1,013**	**3.5**
85 years and older	**116**	**5.4**
Artificial joint		
All ages	**1,294**	**0.5**
65 years and older	**804**	**2.8**
85 years and older	**79**	**3.7**

[1] *Includes all types of implants reported, such as artificial joints, fixation devices, artificial heart valves, intraocular lens implants, pacemakers, ear vent tubes, infusion pumps, dental implants, silicone implants, and artificial veins and arteries.*

[2] *Number of fixation device implants refers to the number of body sites containing the devices (e.g. pins, screws, wires, plates, or rods) that were implanted. It is not the actual number of devices in a particular site.*

Source: *Moss, AJ; Hamburger, S; Moore, RM, et al. Use of selected medical device implants in the United States, 1988. Advance Data from Vital and Health Statistics; No. 191, Hyattsville, Md: National Center for Health Statistics 1991.*

Fixation devices are used in osteosynthesis procedures and primarily involve the surgical treatment of fractures. In these procedures, most commonly known as open reduction and internal fixation, the surgeon implants devices such as nails, pins, rods, screws, bands or metal plates to correct structural defects.

Artificial joints are a biomechanical rather than biologic solution to severe joint disease and fracture. These implants are used in patients with substantial pain or marked functional disabilities related to advanced osteoarthritis or rheumatoid arthritis, fractures, and bone disorders such as aseptic necrosis for whom less extensive procedures have not been successful or are not medically feasible.

Fixation devices were reported more frequently than any other type of implant. Over 4.4 million persons, 1.8% of the population, reported having at least one fixation device. A greater proportion of the older population reported having a fixation device, with 3.5% of those age 65 and older and 5.4% of those age 85 and older reporting a fixation device.

Artificial joints were reported less frequently, by approximately 1.3 million persons (0.5% of the population). The probability of having an artificial joint implanted increases with age. Approximately 2.8% of persons 65 and older and 3.7% of those 85 and older reported having an artificial joint.

Although the percentage of the population with a fixation device is higher among the older age groups, the majority of fixation devices are reported by persons under age 65 (Table 2). Almost half (48.0%) of per-

Table 2: Persons with Selected Types of Medical Device Implants by Age and Gender: United States, 1988

	Number of persons (thousands)	Percent distribution
Fixation device[1]		
Age		
Less than 45 years	2,105	48.0
45-64 years	1,264	28.9
65 years and older	1,013	23.1
65-74 years	564	12.9
75-84 years	332	7.6
85 years and older	116	2.7
Gender		
Male	2,543	58.0
Female	1,839	42.0
Artificial joint		
Age		
Less than 45 years	180	13.9
45-64 years	311	24.0
65 years and older	804	62.1
65-74 years	399	30.8
75-84 years	326	25.2
85 years and older	79	6.1
Gender		
Male	546	42.2
Female	749	57.9

[1] Number of fixation device implants refers to the number of body sites containing the devices (e.g. pins, screws, wires, plates, or rods) that were implanted. It is not the actual number of devices in a particular site.

Source: *Moss, AJ; Hamburger, S; Moore, RM, et al. Use of selected medical device implants in the United States, 1988. Advance Data from Vital and Health Statistics; No. 191, Hyattsville, Md: National Center for Health Statistics 1991.*

sons reporting a fixation device are under age 45 and an additional 28.9% are between ages 45 and 64. In contrast, the majority of persons reporting artificial joints (62.1%) are age 65 and older. The divergent age concentration of fixation devices and artificial joints is apparent for both men and women (Figure 1).

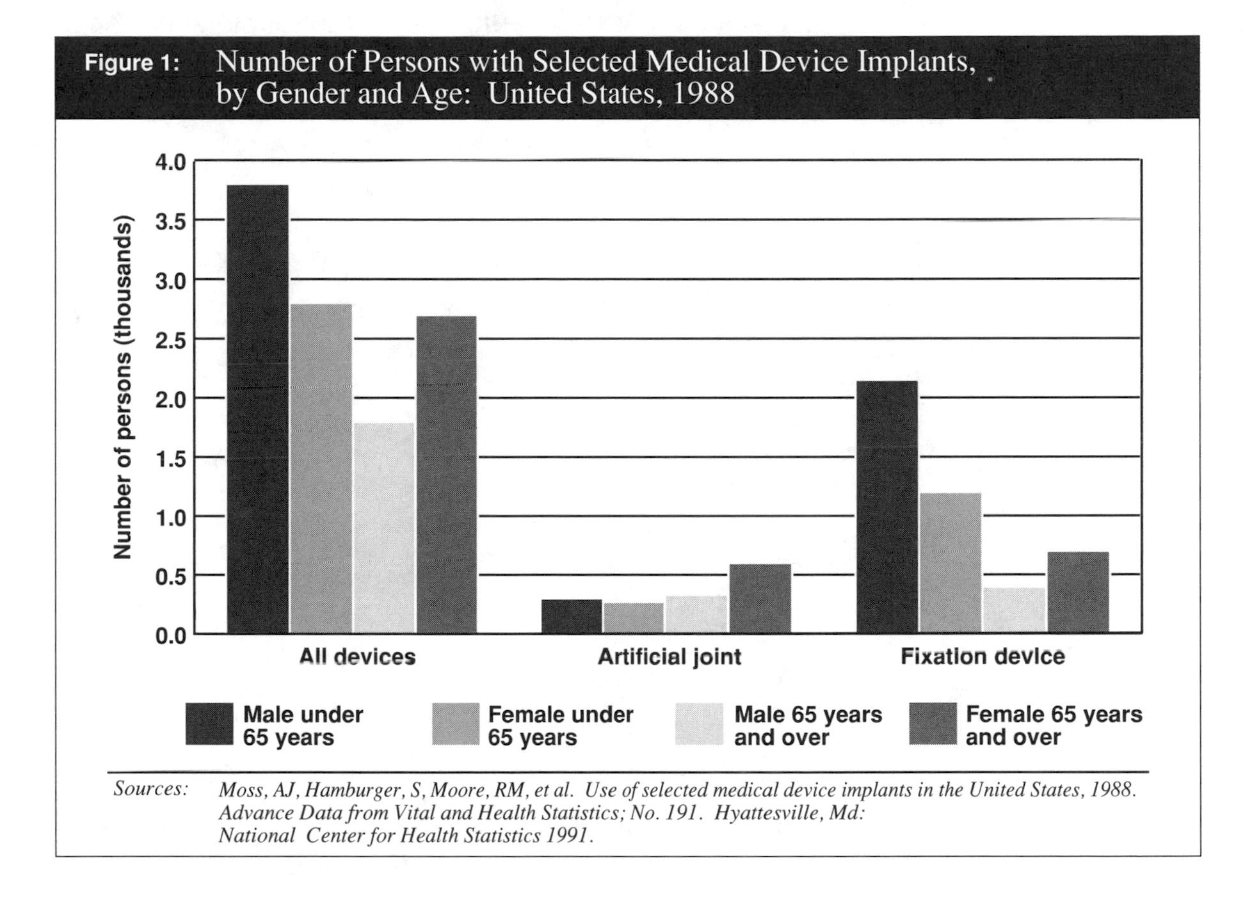

Figure 1: Number of Persons with Selected Medical Device Implants, by Gender and Age: United States, 1988

Sources: *Moss, AJ, Hamburger, S, Moore, RM, et al. Use of selected medical device implants in the United States, 1988. Advance Data from Vital and Health Statistics; No. 191. Hyattesville, Md: National Center for Health Statistics 1991.*

The majority of fixation devices (55.0%) are located in the lower extremities (Table 3). Smaller proportions are located in the upper extremities (13.2%) or torso (11.5%). Artificial joints are located primarily in the hip (50.2%) or knee (32.1%).

The distribution by gender differs between the two types of musculoskeletal implants. In contrast to joint replacements, the majority of fixation devices at all anatomic sites are reported by men. In large part, this reflects two factors: (1) the age of persons receiving each type of implant and (2) the reason for the implant.

As noted above (Table 2), a large majority of persons reporting fixation devices are under age 65 (76.9%) with almost half (48.0%) reported by

Table 3: Fixation Devices and Artificial Joints by Anatomic Site: United States, 1988

	Number of implants (thousands)	Percent distribution				
		Total	Male	Female	Under 65	65 & older
All fixation devices[1,2]	**4,890**	**100.0**	**57.2**	**42.8**	**76.6**	**23.4**
Head	**351**	**100.0**	**57.3**	**42.7**	**94.6**	***5.4**
Torso	**563**	**100.0**	**62.7**	**32.3**	**86.7**	**13.3**
Upper extremity	**646**	**100.0**	**70.4**	**29.6**	**85.0**	**15.0**
Lower extremity	**2,690**	**100.0**	**52.8**	**47.2**	**70.6**	**29.4**
Other site	**622**	**100.0**	**57.9**	**42.1**	**74.9**	**25.1**
All artificial joints	**1,625**	**100.0**	**39.9**	**60.1**	**32.9**	**61.7**
Hip joint	**816**	**100.0**	**37.5**	**62.5**	**33.3**	**66.7**
Knee joint	**521**	**100.0**	**41.8**	**58.2**	**31.5**	**68.5**

**Figure does not meet standards of reliability or precision.*
[1]Each fixation device represents a single body site, regardless of the number of pins, screws, wires, plates, rods, clips or nails that were used to hold or fasten it in a fixed position.
[2]Includes all known and unknown sites.

Source: *Moss, AJ; Hamburger, S; Moore, RM, et al. Use of selected medical device implants in the United States, 1988. Advance Data from Vital and Health Statistics; No. 191, Hyattsville, Md: National Center for Health Statistics 1991.*

persons under age 45. The United States population under age 45 is more evenly divided by gender (49.6% male) than is the population over age 65 (40.7% male). Other factors being equal, this would increase the percentage of fixation devices among males. Secondly, the most frequently reported reason for implanting a fixation device was injury (68.8%) (Table 4). Approximately 57% of musculoskeletal injuries reported each year in the United States occur among men. In contrast, the primary indication for an artificial joint was arthritis (47.9%), which is most prevalent among persons 65 and older.

Also a function of the younger age at which they are implanted is the length of time fixation devices are in use; 62.2% of fixation devices have been in use at least 5 years (Table 5), as compared with 48.6% of artificial joints that have been in use for the same length of time.

Overall, and among each age group, more than 90% of both fixation devices and artificial joints currently in use have never been replaced (Table 6). Fixation devices were most likely to have been replaced

Table 4: Fixation Devices and Artificial Joints
Reasons for Original Implant: United States, 1988

	Number of implants (thousands)	Percent distribution
All fixation devices[1]	4,890	100.0
Injury	3,362	68.8
Deformity	275	5.6
Cancer	*39	*0.8
Infection	*21	*0.4
Other	1,140	23.3
All artificial joints[1]	1,625	100.0
Arthritis	778	47.9
Osteoarthritis	246	15.1
Rheumatoid arthritis	190	11.7
Injury	460	28.3
Pain	135	8.3
Other	373	23.0

[1] *More than one reason could be indicated.*
**Figure does not meet standard of reliability or precision.*

Source: *Moss, AJ; Hamburger, S; Moore, RM, et al. Use of selected medical device implants in the United States, 1988. Advance Data from Vital and Health Statistics; No. 191, Hyattsville, Md: National Center for Health Statistics 1991.*

Table 5: Fixation Devices and Artificial Joints
Length of Time in Use of Current Device: United States, 1988

	Number of devices	*Percent distribution*			
	(Thousands)	Total	Less than 1 year	1-4 years	5 years & older
All fixation devices	4,890	100.0	12.9	24.9	62.2
Age					
Less than 18 years	193	100.0	30.8	43.4	25.3
18-44 years	2,168	100.0	12.6	23.6	63.8
45-64 years	1,388	100.0	10.3	21.8	67.8
65 years and older	1,142	100.0	13.4	27.9	58.7
Gender					
Male	2,799	100.0	11.3	22.7	66.0
Female	2,092	100.0	15.0	27.8	57.2
All artificial joints	1,625	100.0	16.6	34.8	48.6
Age					
Less than 45 years	222	100.0	*5.6	31.0	63.5
45-64 years	400	100.0	15.8	40.6	43.3
65 years and older	1,003	100.0	19.3	33.3	47.4
Gender					
Male	649	100.0	15.6	34.9	49.5
Female	976	100.0	17.3	34.8	48.0

**Figure does not meet standards of reliability or precision.*

Source: *Moss, AJ; Hamburger, S; Moore, RM, et al. Use of selected medical device implants in the United States, 1988. Advance Data from Vital and Health Statistics; No. 191, Hyattsville, Md: National Center for Health Statistics 1991.*

Table 6: Fixation Devices and Artificial Joints, Percentage of Devices Never Replaced by Age Category of Recipient: United States, 1988

	(Percent)	
	Fixation devices	Artificial joints
Total	**94.6**	**92.2**
Age		
Less than 18 years	**90.6**	***100.0**
18-44 years	**93.3**	**93.1**
45-64 years	**94.9**	**90.9**
65 years and older	**97.2**	**92.5**

**Figure does not meet standard of reliability or precision.*

Source: *Moss, AJ; Hamburger, S; Moore, RM, et al. Use of selected medical device implants in the United States, 1988. Advance Data from Vital and Health Statistics; No. 191, Hyattsville, Md: National Center for Health Statistics 1991.*

among those under age 18 and least likely among those 65 and older. Recipients in the 45 to 64 age group were most likely to have had an artificial joint replaced. Among those 65 and older, who have the majority of artificial joints, 92.5% have not had the artificial joint replaced.

Approximately 33.2% of those with a fixation device reported having a problem with it (Table 7). Problems were most likely to be reported by those age 18 to 44. A slightly smaller percentage, 31.6%, reported having a problem with their artificial joint. Those in the 18 to 44 age group were most likely to report a problem (46.6%); among those age 65 and older only 28.2% reported a problem.

Table 7: Fixation Devices and Artificial Joints, Percentage of Devices with One Problem or More Indicated, by Age Category of Recipient: United States, 1988

	(Percent)	
	Fixation devices	Artificial joints
Total	**33.2**	**31.6**
Age		
Less than 18 years	**22.3**	***33.3**
18-44 years	**36.1**	**46.6**
45-64 years	**33.7**	**32.4**
65 years and older	**29.1**	**28.2**

**Figure does not meet standard of reliability or precision.*

Source: *Moss, AJ; Hamburger, S; Moore, RM, et al. Use of selected medical device implants in the United States, 1988. Advance Data from Vital and Health Statistics; No. 191, Hyattsville, Md: National Center for Health Statistics 1991.*

Major Joint Procedures

Joint procedures, including total joint replacement, result in a substantial use of health care resources. On average, over 531,000 arthroplasties were performed annually in the United States from 1985 through 1988 in short-term general hospitals. Almost three-fourths of these were hip (38.4%) or knee procedures (35.8%) (Figure 2).

Figure 2: Arthroplasty Procedures by Anatomic Site Annual Average: United States, 1985-1988

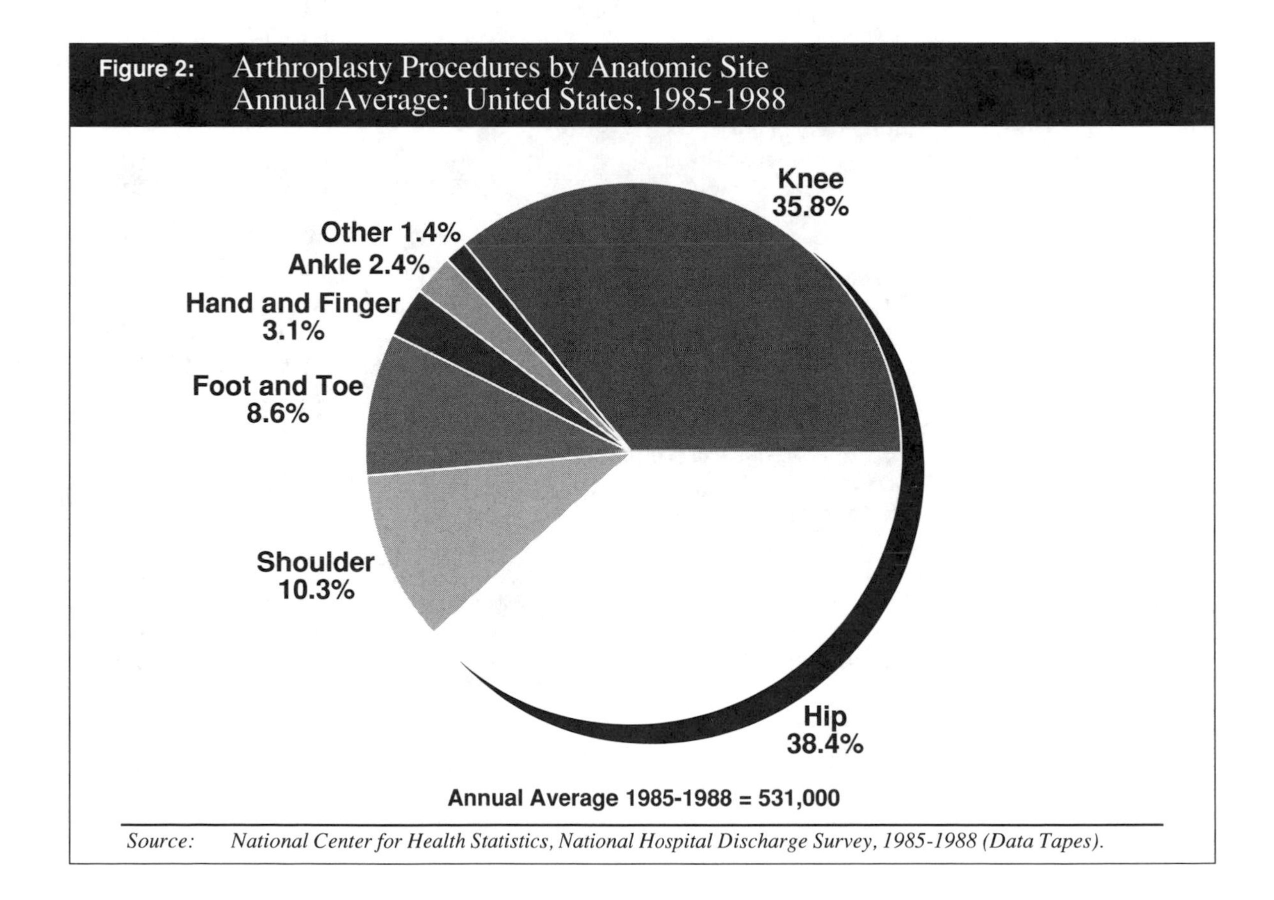

Source: National Center for Health Statistics, National Hospital Discharge Survey, 1985-1988 (Data Tapes).

Approximately 204,000 hip arthroplasties were performed on average each year (Table 8). These included 123,000 total hip replacements and 58,000 replacements of the head of the femur. An average of 190,000 knee arthroplasties were also performed with total knee replacements accounting for approximately half of these (95,000).

Table 8: Arthroplasty and Total Joint Replacement Procedures: United States, 1985-1988, Annual Average

Description	ICD-9-CM Code	Average	Percent
Arthroplasty of foot & toe			
Arthroplasty of foot & toe with synthetic prosthesis	81.31	11,000	2.1
Other arthroplasty of foot & toe	81.39	35,000	6.5
Total, Arthroplasty of foot & toe		46,000	8.6
Arthroplasty of knee			
Total knee replacement	81.41	95,000	17.9
Triad knee repair	81.43	*	-
Patellar stabilization	81.44	4,000	0.7
Other repair of cruciate ligaments	81.45	36,000	6.8
Other repair of collateral ligaments	81.46	14,000	2.6
Other repair of knee	81.47	41,000	7.7
Total, Arthroplasty of knee		190,000	35.8
Arthroplasty of ankle			
Total ankle replacement	81.48	*	-
Other repair of ankle	81.49	13,000	2.4
Total, Arthroplasty of ankle		13,000	2.4
Arthroplasty of hip			
Total hip replacement with methyl methacrylate	81.51	35,000	6.5
Other total hip replacement	81.59	88,000	16.6
Replacement of head of femur with methyl methacrylate	81.61	10,000	1.9
Other replacement of head of femur	81.62	48,000	9.1
Replacement of acetabulum with methyl methacrylate	81.63	*	-
Other replacement of acetabulum	81.64	2,000	0.5
Other repair of hip	81.69	19,000	3.5
Total, Arthroplasty of hip		204,000	38.4
Arthroplasty of hand & finger			
Arthroplasty of hand & finger with synthetic prosthesis	81.71	7,000	1.3
Other repair of hand & finger	81.79	9,000	1.8
Total, Arthroplasty of hand & finger		16,000	3.1
Arthroplasty of upper extremity, except hand			
Arthroplasty of shoulder with synthetic prosthetis	81.81	3,000	0.6
Repair of recurrent dislocation of shoulder	81.82	16,000	3.1
Other repair of shoulder	81.83	35,000	6.6
Arthroplasty of elbow with synthetic prosthetsis	81.84	*	-
Other repair of elbow	81.85	*	-
Arthroplasty of carpals with synthetic prosthesis	81.86	2,000	0.5
Other repair of wrist	81.87	3,000	0.5
Total, Arthroplasty of upper extremity, except hand		62,000	11.8
Total, Arthroplasty		531,000	100.0

** Figure does not meet standard of reliability or precision. (Estimates of procedures that are less than 10,000 are based on a relatively small number of actual discharge records. Estimates of procedures that are less than 5,000 are based on fewer than 30 records. If the relative standard error is larger than 30%, the estimate is not reported. Also note that, due to rounding, totals may not reflect the sum of the procedures listed above.*

Source: *National Center for Health Statistics, National Hospital Discharge Survey, 1985-88 (Data Tapes).*

Joint replacements

The number of joint replacements, primarily total knee and total hip replacements, has increased substantially in recent years. The annual number of total knee and total hip replacements provided to the medicare population between 1985 and 1989 is indicated in Figure 3. Medicare enrollees account for approximately 70% of major joint replacements. Total knee replacements increased from 45,368 in 1985 to 80,647 in 1989, an increase of 77.8%. Over the same period, total hip replacements increased 23.9% from 50,799 to 62,918. The number of knee replacements increased at a 15.5% annual rate compared with 5.5% for total hip replacements (compounded annual rate).

Figure 3: Total Hip and Knee Replacements, Medicare Procedures* United States, 1985-1989

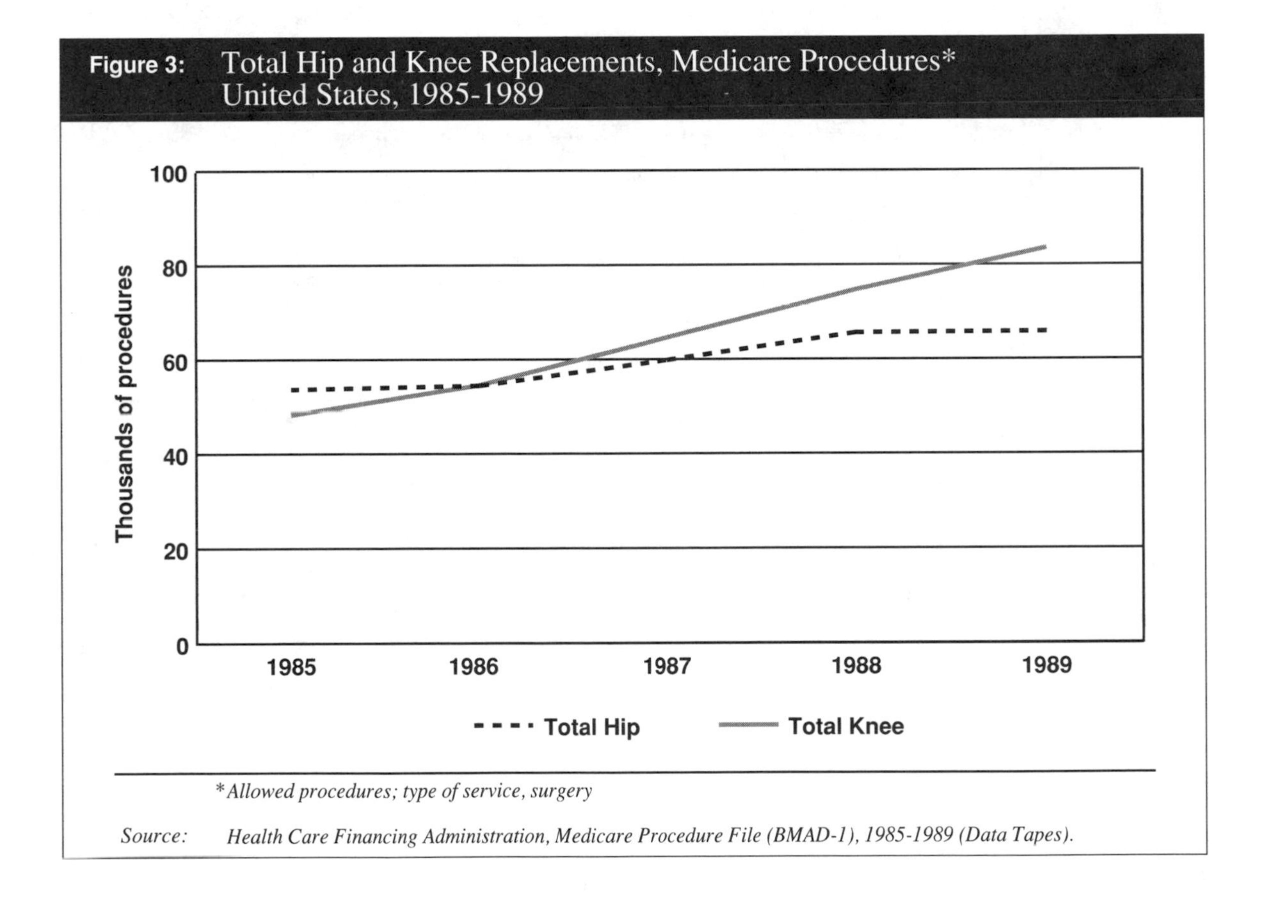

**Allowed procedures; type of service, surgery*

Source: *Health Care Financing Administration, Medicare Procedure File (BMAD-1), 1985-1989 (Data Tapes).*

Total joint replacements and other major joint procedures have a substantial impact on medicare utilization and charges. Total knee replacements and total hip replacements rank fourth and fifth among surgical procedures in Medicare Part B allowed charges. Allowed charges for these procedures totaled almost $327 million in 1989 and accounted for approximately 4% of medicare allowed charges for surgeries. The frequency and allowed charges for these and selected medicare major joint procedures are indicated in Table 9.

Table 9: Frequency and Allowed Charges for Selected Major Joint Procedures: 1989, Medicare Population

CPT Code		Frequency	Allowed charges (thousands)
27447	Total knee replacement	**80,647**	**$180,954**
27486	Revision of total knee, arthroplasty, one component	**1,804**	**3,559**
27487	Revision of total knee, arthroplasty, all components	**4,323**	**11,524**
27130	Total hip replacement	**62,918**	**145,791**
27132	Conversion of previous hip surgery to total hip replacement	**4,171**	**10,043**
27134	Revision of total hip replacement, both components	**7,817**	**23,480**
27137	Revision of total hip replacement, acetabular component only	**1,999**	**5,024**
27138	Revision of total hip replacement, femoral component only	**2,344**	**5,670**
27125	Hemiarthroplasty of hip; prosthesis	**25,873**	**41,847**
27236	Open treatment of closed or open femoral fracture, proximal end, neck, internal fixation or prosthetic replacement	**63,970**	**79,591**
23470	Arthroplasty with proximal humeral implant	**2,987**	**3,731**
23472	Arthroplasty with glenoid and proximal humeral replacement	**2,107**	**4,631**

Source: *Health Care Financing Administration, Medicare procedure File, BMAD-1, 1989.*

Principal diagnoses by procedure

The leading principal diagnoses associated with four major joint procedures, total hip replacement, total knee replacement, total shoulder replacement, and femoral head replacement, are indicated in Tables 10 through 13*. The most frequent principal diagnosis associated with the three total joint replacement procedures was osteoarthrosis and related disorders.

For total hip replacements, osteoarthrosis and allied disorders was the principal diagnosis for 55.4% of these procedures (Table 10). Fracture of the neck of the femur accounted for 18.3% and mechanical complications of an implant accounted for 13.7% of these procedures. Smaller percentages resulted from bone disorders, such as aseptic necrosis of bone and nonunion/malunion of fractures, and from arthropathies.

*Data contained in Tables 10 through 16 were obtained from Consolidated Consulting Group's (CCG) Medicare Hospital Utilization Data Bases, which link Medicare Part A and Part B data.

Table 10: Leading Principal Diagnoses for Total Hip Replacements Medicare Hospital Inpatients, 1989

Diagnosis	Percent of total cases
Osteoarthrosis and allied disorders	54.4
Fracture of neck of femur	18.3
Mechanical complication of implant	13.7
Aseptic necrosis of bone, nonunion/malunion of fracture, pathological fractures, and other bone disorders	6.9
Rheumatoid arthritis and other inflammatory polyarthropathies	2.2
Unspecified arthropathies	1.1
All other	3.3

Notes: *A principal diagnosis is the condition established after study to be chiefly responsible for causing a hospitalization or use of hospital services.*

Source: *CCG's Medicare Hospital Utilization Data Bases*

The distribution of principal diagnoses associated with total knee replacements is more concentrated, with 85.4% associated with osteoarthrosis (Table 11). The next most frequent principal diagnosis was mechanical complication of an implant, followed by rheumatoid arthritis and other inflammatory polyarthropathies (4.9%).

Table 11: Leading Principal Diagnoses for Total Knee Replacements Medicare Hospital Inpatients, 1989

Diagnosis	Percent of total cases
Osteoarthrosis and allied disorders	85.4
Mechanical complication of implant	5.7
Rheumatoid arthritis and other inflammatory polyarthropathies	4.9
Unspecified arthropathies	1.7
All other	2.2

Notes: *A principal diagnosis is the condition established after study to be chiefly responsible for causing a hospitalization or use of hospital services.*

Source: *CCG's Medicare Hospital Utilization Data Bases*

For total shoulder replacements, osteoarthrosis and allied disorders was listed as the diagnosis in 33.7% of the cases, and fractures of the humerus were listed as the principal diagnosis in approximately 33.6% of cases (Table 12). Bone disorders accounted for 9.6% and inflammatory polyarthropathies 8.1% of principal diagnoses.

Table 12: Leading Principal Diagnoses for Total Shoulder Replacements Medicare Hospital Inpatients, 1989

Diagnosis	Percent of total cases
Osteoarthrosis and allied disorders	**33.7**
Fracture of humerus	**33.6**
Aseptic necrosis of bone, nonunion/malunion of fracture, pathological fractures, and other bone disorders	**9.6**
Rheumatoid arthritis and other inflammatory polyarthropathies	**8.1**
Unspecified arthropathies	**2.5**
Mechanical complication of implant	**2.1**
All other	**10.1**

Notes: *A principal diagnosis is the condition established after study to be chiefly responsible for causing a hospitalization or use of hospital services.*

Source: *CCG's Medicare Hospital Utilization Data Bases*

The leading principal diagnosis for femoral head replacements was fracture of the neck of the femur, which accounted for 87.6% of cases (Table 13). Aseptic necrosis and other bone disorders accounted for 3.9% of cases and mechanical complications of implants, 2.8%. In contrast to total joint replacement, relatively few femoral head replacements were associated with osteoarthrosis and related disorders.

Table 13: Leading Principal Diagnoses for Femoral Head Replacements Medicare Hospital Inpatients, 1989

Diagnosis	Percent of total cases
Fracture of neck of femur	**87.6**
Aseptic necrosis of bone, nonunion/malunion of fracture, pathological fractures, and other bone disorders	**3.9**
Mechanical complication of implant	**2.8**
Osteoarthrosis and allied disorders	**2.3**
All other	**3.4**

Notes: *A principal diagnosis is the condition established after study to be chiefly responsible for causing a hospitalization or use of hospital services.*

Source: *CCG's Medicare Hospital Utilization Data Bases*

Length of stay

The average lengths of stay (ALOS) for medicare patients undergoing total joint or femoral head replacements are indicated in Table 14. The longest length of stay, 13.6 days, was associated with femoral head replacements. Total hip replacements resulted in ALOS of 12.1 days and total knee replacements 10.9 days. Only total shoulder replacement had an ALOS (7.7 days) that was less than the average for medicare surgical patients (8.8 days).

Table 14: Average Lengths of Stay for Medicare Hospital Inpatients Receiving Major Joint Procedures, 1989

Procedure	Average length of stay (days)
Total hip replacement	12.1
Femoral head replacement	13.6
Total knee replacement	10.9
Total shoulder replacement	7.7
All Medicare Inpatients	**8.8**

Notes: *The calculations for joint replacement patients were restricted to non-trauma patients receiving single joint replacements, i.e. those falling into DRG 209 for hip and knee replacements or DRG 223 for total shoulder replacements. Patients receiving multiple or bilateral joint replacements and/or in trauma had significantly higher lengths of stay.*

The average length of stay for all Medicare inpatients is restricted to Medicare inpatients at short-stay, acute care nonfederal hospitals.

Source: *CCG's Medicare Hospital Utilization Data Bases*

One factor that may contribute to the comparatively long ALOS associated with femoral head replacement was the advanced age of the patient. The majority of persons who underwent a femoral head replacement (57.5%) were age 80 and older compared with only 25.3% of persons undergoing total hip replacement (Table 15). Only 21.0% of those undergoing total shoulder replacement were age 80 and older which is approximately the same percentage as that of the medicare population 80 and older (20.9%). Total knee replacement, with 16.3% of cases, has the smallest percent of cases age 80 and older.

Discharge destination

Age may also be a factor in relation to the discharge destination associated with each of the four procedures (Table 16). Persons having undergone a femoral head replacement were least likely to be listed as

Table 15: Medicare Hospital Inpatients Receiving Major Joint Procedures 1989, Percent Distribution by Age Group

Procedure	Under 65 years	65-69	70-74	75-79	80-84	85-89	90 years & older
Total hip replacement	6.2	22.1	24.4	22.1	15.0	7.2	3.1
Femoral head replacement	2.0	7.7	12.9	19.9	23.9	20.3	13.3
Total knee replacement	5.3	26.3	28.9	23.3	12.2	3.6	0.5
Total shoulder replacement	9.3	22.9	24.6	22.2	13.4	6.2	1.4
Medicare Population	9.6	28.8	23.1	17.6	11.4	*9.5	

** This percentage is for all Medicare enrollees aged 85 or older. Enrollment figures for individuals aged 90 and over in 1989 were not available at the time of this publication.*

Source: *CCG's Medicare Hospital Utilization Data Bases*

Table 16: Discharge Destinations for Medicare Patients Undergoing Major Joint Procedures, 1989

	(Percent)				
Procedure	Home	SNF	HHS	ICF	Other
Total hip replacement	57.4	15.9	14.7	2.3	9.7
Femoral head replacement	32.2	35.3	10.1	7.9	14.5
Total knee replacement	67.4	8.9	16.0	1.0	6.7
Total shoulder replacement	75.4	8.3	10.7	2.2	3.4

Notes: *SNF is an acronym for skilled nursing facility.*

HHS is an acronym for home health services; patients with this discharge destination were discharged to their personal home under the care of a certified home health service organization.

ICF is an acronym for intermediate care facility.

"Other" destinations include deaths, transfers to other hospitals and unspecified destinations.

The percentages of patients discharged to personal homes with and without any home health care service arrangements may be midleading. A number of individuals discharged to personal homes subsequently require and receive home health services.

Source: *CCG's Medicare Hospital Utilization Data Bases*

discharged to home. These persons were the most likely to be discharged to a skilled nursing facility (35.3% of persons) or intermediate care facility (7.9%). Persons undergoing a shoulder replacement or knee replacement were most likely to be discharged to home.

References

Consolidated Consulting Group. Medicare Hospital Utilization Data Bases. Fairfax, VA.

Health Care Financing Administration. Medicare Procedure File (BMAD-1), 1985-1989 (Data tapes).

Moss AJ, Hamburger S, RM, et al. Use of Selected Medical Device Implants in the United States, 1988. Advance Data from Vital and Health Statistics; No. 191, Hyattsville, Md: National Center for Health Statistics 1991.

National Center for Health Statistics. National Hospital Discharge Survey, 1985-1988 (Data tapes).

Section 5

Costs of Musculoskeletal Conditions

Costs of Musculoskeletal Conditions

Introduction

The impact of musculoskeletal conditions in the United States includes the medical resources used for care, treatment, and rehabilitation; reduced or lost productivity; and the pain and suffering of patients, their families and friends. These conditions impose a substantial burden on the individual and on society as a whole. This burden must be translated into economic terms to better understand its magnitude compared with that of other major chronic illnesses so that informed decisions about health care policy can be made (Rice, et al, 1985).

This chapter summarizes the methods and sources of data used in estimating the direct and indirect costs of musculoskeletal conditions and presents the results.

Methods and data sources

Cost-of-illness studies are typically divided into two major categories: direct costs and indirect costs. Direct costs are those for which payments are made, and indirect costs are those for which resources are lost. Indirect costs consist of morbidity costs, the value of lost productivity by persons unable to perform their usual activities or to perform them at a level of full effectiveness because of illness; and mortality costs, the value of lost productivity because of premature death caused by the illness. Lost productivity is calculated as the present discounted value of future market earnings, as well as an imputed value for housekeeping services (Hodgson and Meiners, 1982). Table 1 indicates the present value of lifetime earnings by age and gender for 1988.

To ascertain the cost of musculoskeletal conditions, the human capital approach was used in this study. According to this approach, a person produces a stream of output that is valued at market earnings or the imputed value of housekeeping services. The approach is based on a social perspective and has the advantage of using reliable and readily available data. The human capital approach yields low values for children and the retired elderly because the value of human life is based on market earnings. Therefore, the cost of musculoskeletal conditions presented here may be understated, because a relatively large number of elderly persons suffer from these conditions.

Table 1: Present Value of Lifetime Earnings[1] by Age and Gender, 1988

Age	Males	Females
Under 1 year	**$479,631**	**$370,124**
1-4 years	**517,541**	**399,180**
5-9 years	**591,409**	**455,956**
10-14 years	**685,471**	**528,293**
15-19 years	**778,168**	**591,192**
20-24 years	**838,892**	**616,449**
25-29 years	**850,049**	**601,885**
30-34 years	**817,389**	**560,543**
35-39 years	**748,341**	**502,760**
40-44 years	**641,025**	**429,877**
45-49 years	**510,281**	**347,672**
50-54 years	**372,125**	**265,548**
55-59 years	**234,976**	**188,017**
60-64 years	**124,077**	**120,798**
65-69 years	**57,827**	**70,080**
70-74 years	**27,388**	**37,685**
75-79 years	**13,358**	**19,031**
80-84 years	**6,647**	**9,151**
85 years & over	**2,041**	**2,326**

[1]Discounted at 4%.

One major category of costs is omitted here—that of pain and suffering. No one has successfully quantified this dimension of illness, yet some diseases, such as musculoskeletal conditions, impose more suffering than others.

Two other categories of cost are purposefully omitted—transfer payments and taxes. When income loss is used as a measure of indirect costs, adding pension or relief payments would be double counting. As for tax payments, it would be double counting to add income tax losses to loss of earnings and triple counting if the tax receipts were added to public payments for medical care. Transfer and tax payments may alter the allocation of resources, but are not themselves a use of resources.

Also, in this study, prevalence-based costs were used to estimate the cost of musculoskeletal conditions. This method provides an estimate of the direct and indirect economic burden incurred during the base period as a result of the prevalence of illness, usually one year. Included is the cost of base-year manifestations or associated disability with onset in the base year or at any time prior to that year. This study employs 1988 as the base year.

Definition of musculoskeletal conditions

The term "musculoskeletal conditions" comprises any of the following diagnoses: Musculoskeletal diseases and connective tissue disorders; fractures, dislocations and sprains, open wounds and crushing injuries, traumatic amputations, and other selected injuries affecting the musculoskeletal system; neoplasms of bone, connective tissues, selected lymphomas; selected congenital musculoskeletal deformities and anomalies; and other selected diseases affecting the musculoskeletal system, based on the International Classification of Diseases Ninth Revision, Clinical Modification (ICD•9•CM).

The estimation of the economic costs of musculoskeletal conditions is complex, involving a variety of methods and sources of data that are described below.

Direct costs

In general, direct costs (medical expenditures) are estimated as the product of two components: number of services and unit prices or charges. The following sections include a description of the sources of data and the methods used for estimating the costs for each type of medical expenditure.

Hospital inpatient costs

The number of inpatient days for persons hospitalized with primary diagnoses of musculoskeletal conditions by age and sex, as reported in the 1988 National Hospital Discharge Survey Public Use Tapes, is multiplied by average expenses per patient day in community hospitals (American Hospital Association, 1991). The amount for 1988 ($631.91) was adjusted upward to reflect the severity and higher costs associated with care of patients who have musculoskeletal conditions based on data reported by Medicare in terms of diagnosis-related groups (DRGs) (Latta and Helbing, 1991). The ratio of average program payments weighted for frequency for all DRGs listed under the major diagnostic category, diseases and disorders of the musculoskeletal system and connective tissues, to all hospital payments (1.43) was applied to the average expense per patient day for all community hospitals.

Hospital outpatient services

Recent advances in medical technology allow hospitals to provide increased services on an outpatient basis. Outpatient services include ambulatory surgery, physical therapy, rehabilitation, emergency rooms, and outpatient departments. In 1988, hospitals reported 269 million outpatient visits. Nearly half of all operations performed in community hospitals were done on an outpatient basis (American Hospital Association, 1989).

Unfortunately, data are not available on outpatient services by diagnosis. Therefore, the charges for musculoskeletal services provided on an outpatient basis were estimated from gross revenues reported by the American Hospital Association in its annual report for all inpatient and outpatient services. Revenues for outpatient services were 25.3% of total inpatient revenues. This percentage was applied to inpatient costs to estimate outpatient costs for patients with musculoskeletal conditions.

Physician inpatient services

Included in physician inpatient services are charges for physician services in the hospital, including surgery. Data on the number of surgical procedures performed on the musculoskeletal system by age and sex were obtained from the 1988 National Hospital Discharge Survey data tapes. Medicare-allowed charges in 1988 for the 25 most frequent surgical procedures performed on patients with musculoskeletal conditions were obtained from Medicare data tapes. These charges were applied to the 25 most frequent surgical procedures reported in the National Hospital Discharge Survey, resulting in a mean surgical fee of $1,357 per procedure. These 25 procedures represented 52% of the 3 million total procedures performed in 1988. For the remaining 1.4 million procedures, a conservative charge of $400 was applied.

In addition to inhospital surgical fees, there are also physician visits to patients hospitalized for musculoskeletal conditions. Costs for these visits are based on the number of days of care for hospitalized patients. It is conservatively estimated that half of these visits (12.25 million out of 24.5 million) result in a charge; the other half would be provided by interns and residents who are covered in the costs per patient day of community hospitals. A charge per visit of $37.30 is used, the mean fee for a follow-up hospital visit for orthopaedic surgery reported by the American Medical Association (AMA Center for Health Policy Research, 1989). This mean fee was multiplied by the 12.25 million visits by age and sex and added to the inhospital surgical fees.

Physician outpatient services

The number of office visits by age and sex for persons with musculoskeletal conditions is reported by the National Ambulatory Care Survey conducted by the National Center for Health Statistics (1985). The mean fee of an orthopaedic surgeon in 1988 for an office visit by a new orthopaedic patient was $67.30. It was $34.90 for an established patient (AMA, 1989). Assuming there are two visits by an established patient for every visit by a new patient, a fee of $45.70 was applied to the number of office visits to obtain the amounts spent for physician outpatient services.

Other practitioners' services

This category of expenditures includes payments to private duty nurses, chiropractors, podiatrists, social workers, nurses, physical therapists, clinical psychologists, naturopaths and others. Data on these services are available from the 1980 National Medical Care Utilization and Expenditure Survey. A report on musculoskeletal conditions showed that expenditures for other professional services provided to patients with these conditions were 62.5% of expenditures for physician services (Murt, et al, 1986). This percentage was applied to expenditures for outpatient physician services by age and sex to obtain amounts paid for outpatient practitioners' services.

Drugs

Included in this category are expenditures for medications prescribed for musculoskeletal conditions. Excluded are nonprescription drugs and drugs administered or prescribed in institutional and other settings. The National Medical Care Utilization and Expenditure Survey reported per capita expenditures for prescribed medications by age and sex for musculoskeletal conditions (Murt, et al, 1986). These per capita amounts were inflated to 1988 dollars on the basis of the prescription-drug component of the Consumer Price Index published by the Bureau of Labor Statistics. The 1988 per capita amounts were applied to the estimated number of civilian noninstitutionalized persons with musculoskeletal conditions in 1988 to obtain total expenditures for drugs.

Nursing home costs

Nursing home costs are based on the number of nursing home residents with musculoskeletal conditions reported by the National Nursing Home Survey conducted by the National Center for Health Statistics in 1985. Costs were estimated by multiplying the number of residents by the annual charge in 1985 which was $17,471 (National Center for Health Statistics, 1989). This cost was inflated to 1988 dollars by the CPI medical care services index and totaled $21,438.

Prepayment and administration

This category of expenditure includes the net costs of prepaid insurance and administrative expenses of federal programs. The benefits paid under these programs are not counted as costs of illness because they are payments in which funds are transferred from one payer to another. The net cost to society resulting from this transaction in terms of resources used is zero except for costs incurred in operating the system that effects the transfer. Prepayment and administrative costs are based on the percentage of these costs to total personal health care expenditures, which was 5.8% in 1988, as reported by the Health Care Financing Administration (Lazenby and Letsch, 1989).

Non-health care sector costs

This category of expenditure represents the cost of transportation to and from physicians' offices and other health care sites, extra household help required because of the condition, special diets, retraining and education, and alterations to living quarters. Expenditures were estimated at 15% of total direct health care costs less prepayment and administration costs. They were based on the 1977 estimates of the cost of musculoskeletal conditions as prepared by Holbrook, et al (1984) that relied on research performed by Mushkin and Landerfeld (1978).

Indirect costs

Indirect costs include the value of output lost because of reduced or lost productivity caused by illness, disability, or injury. These costs include the value of lost workdays and housekeeping days and lowered productivity caused by illness or disability, and losses caused by premature death. As noted earlier, indirect costs consist of morbidity and mortality costs.

Morbidity costs

Morbidity costs, the value of reduced or lost productivity, are estimated as the product of (1) the number of persons with musculoskeletal conditions; (2) disability or impairment rates; and (3) average earnings of these persons had they not been affected by these disorders, including an imputed value for housekeeper services. Summing these three products over age, sex and musculoskeletal disorder provides an estimate of the aggregate loss of income among the entire population.

Bed disability days are reported for persons with acute musculoskeletal conditions and for injuries in the National Health Interview Survey (National Center for Health Statistics, 1989). These were converted to bed years; and labor force participation rates, earnings, and an imputed value for housekeeping services were applied by age and sex. Chronic conditions are also reported in the National Health Interview Survey. Because individuals may report more than one chronic condition, the number with musculoskeletal conditions was adjusted downward based on two studies by Yelin and Katz (1990, 1991). For persons 18 to 64 years of age, the proportion reporting inability to work was applied to the number of persons with these conditions (LaPlante, 1991). For the elderly population, the proportion unable to carry on their major activity was applied. Labor force participation rates for the elderly, earnings, and an imputed value for housekeeping services were applied to obtain morbidity costs.

Mortality costs

The estimated mortality costs are the product of the number of deaths and the expected value of an individual's future earnings, with sex and age taken into account. This method of derivation considers life expectancy for different age and sex groups, changing pattern of earnings at successive ages, varying labor force participation rates, an imputed value for housekeeping services, and the appropriate discount rate to convert a stream of earnings into its present worth. A discount rate of 4% was used to convert the stream of lifetime earnings into present value equivalent. An average annual increase of 1% in the future productivity of wage earners was assumed.

Cross-sectional profiles of mean earnings by age and sex as reported by the United States Bureau of the Census (1987) are used and are based on the assumption that the future earnings of the average person will increase with age and experience in accordance with the cross-sectional data for that year.

Marketplace earnings underestimate productivity losses because many persons are not in the labor force. Many of these persons, as well as those in the labor force, perform household services. The value of household work, therefore, must be added to earnings. For this study, estimates are developed of hours spent on household labor, employing regression analysis to control for socioeconomic and demographic factors (Douglass, et al, 1990). The hours are then valued on the basis of 1988 wage rates.

Results

Musculoskeletal conditions represent a wide range of disorders, and the prevalence is high, resulting in widespread use of health care services and considerable costs to society in productivity losses, as summarized below.

Total economic impact

Musculoskeletal conditions imposed a $126 billion burden on the United States economy in 1988 (Table 2). Females accounted for 54% of the total costs and males for the remaining 46% (Tables 3 and 4). The 45 to 64 year old age group accounted for the largest share of costs, 37%, followed closely by the 65 years and over age group, which accounted for 35% of the total (Figure 1), reflecting the high prevalence of musculoskeletal conditions in these age groups.

Table 2: Estimated Cost of All Musculoskeletal Conditions by Age and Type of Cost, 1988

(Cost in millions of dollars)

Type of cost	Total	Under 18 years	18-44 years	45-64 years	65 years & over
Total	**125,962**	**3,579**	**32,464**	**46,481**	**43,438**
Direct costs, total	**60,987**	**3,357**	**13,294**	**12,287**	**32,049**
Hospital inpatient	22,137	1,338	5,568	5,074	10,157
Hospital outpatient	5,600	338	1,409	1,283	2,570
Physician inpatient	3,159	241	952	727	1,239
Physician outpatient	4,008	521	1,579	1,079	829
Other practitioners	2,507	326	987	675	519
Drugs	680	15	145	261	259
Nursing home care	12,391	-	365	1,071	10,955
Prepayment/administration	2,934	161	640	591	1,542
Nonhealth sector	7,571	417	1,650	1,525	3,979
Indirect costs, total	**64,975**	**222**	**19,170**	**34,194**	**11,389**
Morbidity	59,619	-	16,821	32,112	10,686
Mortality[1]	5,356	222	2,349	2,082	703

[1] *Present value of lifetime earnings discounted at 4%.*

Direct treatment and administrative support costs accounted for 48% of the 1988 total; morbidity costs, the value of reduced or lost productivity, 47%; and mortality costs, 4%, based on a 4% discount rate of the value of productivity forgone in future years as a result of premature mortality in 1988 (Figure 2).

Table 3: Estimated Cost of All Musculoskeletal Conditions by Age and Type of Cost: Males, 1988

(Cost in millions of dollars)

Type of cost	Total	Under 18 years	18-44 years	45-64 years	65 years & over
Total	**57,696**	**2,245**	**18,253**	**23,502**	**13,697**
Direct costs, total	**24,853**	**2,119**	**7,826**	**5,283**	**9,625**
Hospital inpatient	**9,797**	**886**	**3,509**	**2,277**	**3,125**
Hospital outpatient	**2,479**	**224**	**888**	**576**	**791**
Physician inpatient	**1,522**	**146**	**611**	**366**	**399**
Physician outpatient	**1,832**	**301**	**794**	**457**	**280**
Other practitioners	**1,145**	**188**	**496**	**286**	**175**
Drugs	**281**	**9**	**73**	**111**	**88**
Nursing home care	**3,516**	**-**	**107**	**300**	**3109**
Prepayment/administration	**1,196**	**102**	**377**	**254**	**463**
Nonhealth sector	**3,085**	**263**	**971**	**656**	**1,195**
Indirect costs, total	**32,843**	**126**	**10,427**	**18,219**	**4,072**
Morbidity	**29,669**	**-**	**8,919**	**16,965**	**3,785**
Mortality[1]	**3,174**	**126**	**1,508**	**1,254**	**287**

[1]*Present value of lifetime earnings discounted at 4%.*

Table 4: Estimated Cost of All Musculoskeletal Conditions by Age and Type of Cost: Females, 1988

(Cost in millions of dollars)

Type of cost	Total	Under 18 years	18-44 years	45-64 years	65 years & over
Total	**68,265**	**1,333**	**14,212**	**22,980**	**29,741**
Direct costs, total	**36,134**	**1,238**	**5,469**	**7,004**	**22,424**
Hospital inpatient	**12,340**	**452**	**2,059**	**2,797**	**7,032**
Hospital outpatient	**3,121**	**114**	**521**	**707**	**1,779**
Physician inpatient	**1,637**	**95**	**341**	**361**	**840**
Physician outpatient	**2,176**	**220**	**785**	**622**	**549**
Other practitioners	**1,362**	**138**	**491**	**389**	**344**
Drugs	**399**	**6**	**72**	**150**	**171**
Nursing home care	**8,875**	**-**	**258**	**771**	**7,846**
Prepayment/administration	**1,738**	**59**	**263**	**337**	**1,979**
Nonhealth sector	**4,486**	**154**	**679**	**869**	**2,784**
Indirect costs, total	**32,131**	**95**	**8,743**	**15,976**	**7,317**
Morbidity	**29,950**	**-**	**7,902**	**15,147**	**6,901**
Mortality[1]	**2,181**	**95**	**841**	**829**	**416**

[1]*Present value of lifetime earnings discounted at 4%.*

Figure 2: Total Cost of All Musculoskeletal Conditions by Type of Cost, 1988

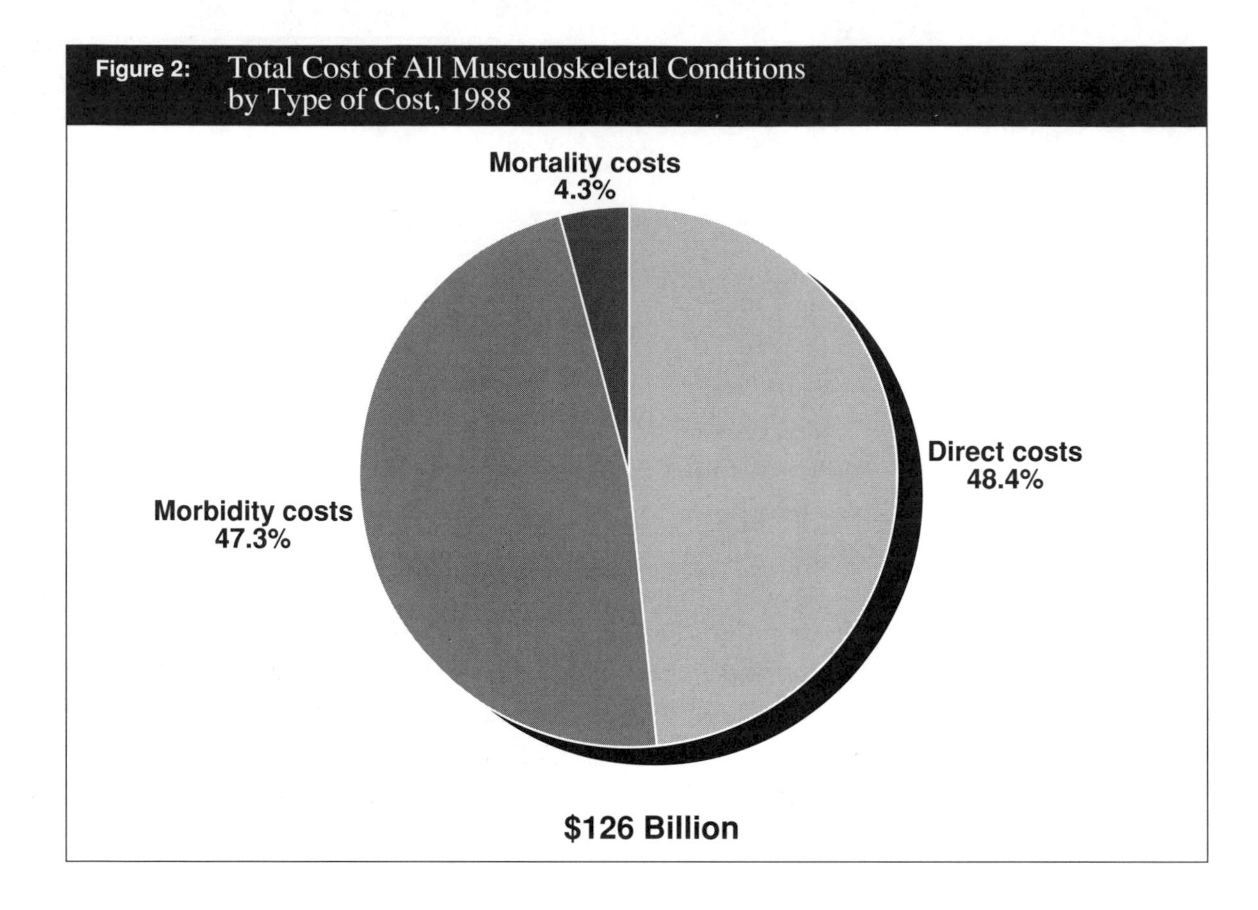

Direct costs

Direct costs for persons suffering from musculoskeletal conditions totaled $61 billion in 1988 or 12.7% of total personal health care spending for all illnesses in that year.

About 36% of the direct costs, $22.1 billion, are expenditures for hospital inpatient care and an additional $5.6 billion are for hospital outpatient care, including ambulatory surgery, physical therapy, rehabilitation, emergency rooms, and outpatient departments. Thus, hospital inpatient and outpatient care totaled $27.7 billion or 45 % of direct costs (Figure 3).

Other treatment costs include $7.2 billion for physician inpatient and outpatient services and $2.5 billion for other professional services, including private duty nurses, chiropractors, podiatrists, social workers, physical and occupational therapists, clinical psychologists, naturopaths and others.

Nursing home expenditures for persons with musculoskeletal conditions amounted to $12.4 billion, 20% of direct costs. Prescription drugs

Figure 3: Direct Cost of All Musculoskeletal Conditions by Type of Service, 1988

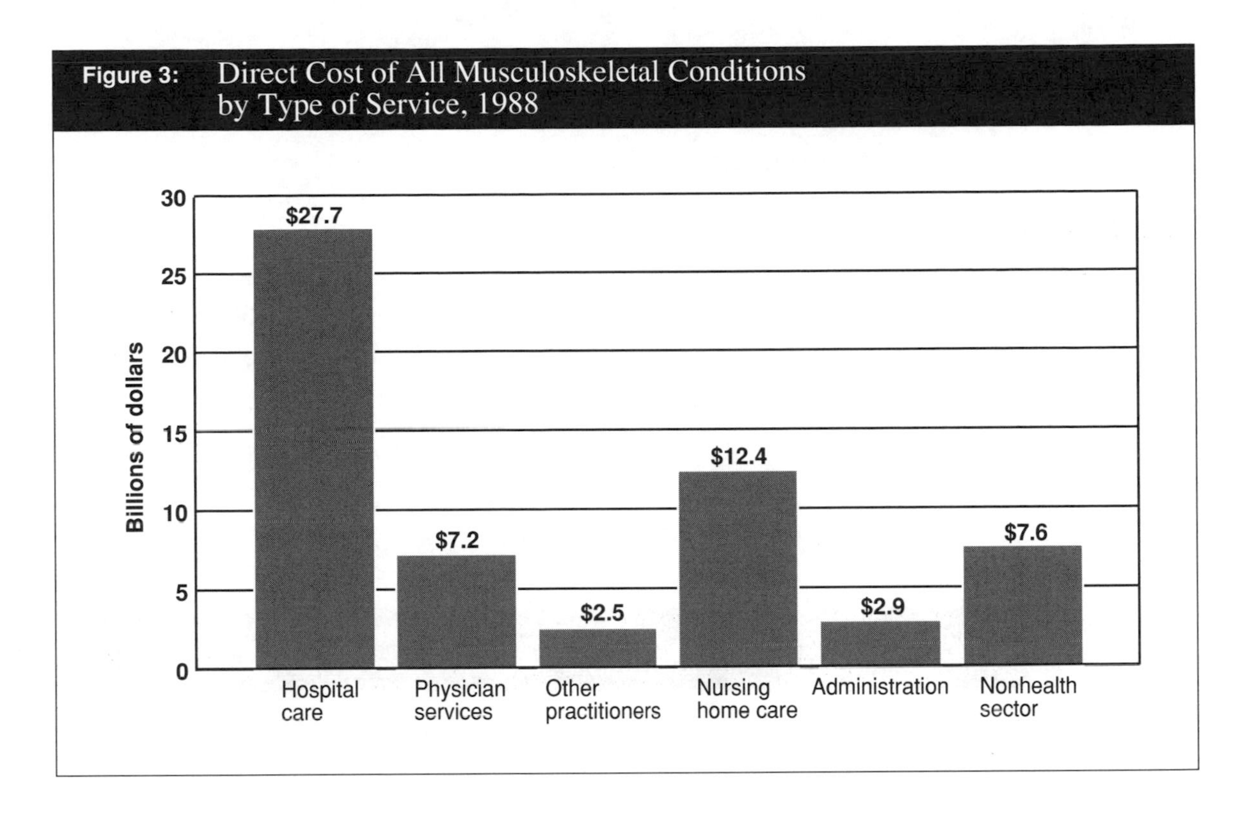

were estimated at $680 million, 1% of the direct costs. The net cost of private insurance and administrative costs is $2.9 billion, 5% of direct costs. Non-health sector costs, including transportation to and from physicians' offices, extra household help, special diets, retraining and eduction, and alterations to living quarters, are estimated at $7.6 billion, 12% of direct costs.

Indirect costs—morbidity

Musculoskeletal morbidity costs, the value of reduced or lost productivity, amounted to $59.6 billion, 92% of the total indirect costs and 47% of total costs. These high morbidity costs reflect the high prevalence of musculoskeletal conditions in the population. For example, the National Health Interview Survey reports a total of 31.2 million persons with arthritic conditions, as well as 2 million with gout, 4.3 million with intervertebral disk disorders, 2.5 million with bone spurs and tendinitis, and 1.5 million with disorders of bone or cartilage. Many persons with these musculoskeletal conditions are not able to work or carry on their major activity. Morbidity costs for women are slightly higher than for men, $30 billion compared with $29.7 billion respectively. Morbidity costs are highest for the 45 to 64 year old age group and account for more than half the total musculoskeletal morbidity costs. The high costs for this age group reflect the high prevalence of these conditions among this population group as well as their high earnings.

Indirect costs—mortality

A total of 44,787 deaths occured as a result of musculoskeletal conditions in 1988. These deaths represent 701,000 million person years lost, or 15.7 years per death and a loss of $5.4 billion to the economy at a 4% discount rate, or $119,582 per death (Table 5). Deaths resulting from musculoskeletal conditions comprised 2.1% of the 2.2 million deaths in the United States in 1988, 2% of the total person years lost, and 2% of the total productivity losses. A detailed breakdown of mortality costs by age and gender is indicated in Tables 6 through 8.

Table 5: Mortality from All Musculoskeletal Conditions: Number of Deaths, Person Years Lost, and Discounted Productivity Losses by Age and Gender, 1988

		Person Years Lost		*Productivity Losses*[1]	
Type of cost	Number of deaths	Number (thousands)	Per death	Amount (millions)	Per death
Both Genders	**44,787**	**701**[2]	**15.7**[2]	**$5,356**[2]	**$119,582**[2]
Under 15 years	**440**	**31**	**70.2**	**222**	**503,470**
15-44 years	**3,585**	**157**	**43.8**	**2,349**	**655,154**
45-64 years	**9,176**	**210**	**22.8**	**2,082**	**226,935**
65 years & over	**31,586**	**304**	**9.6**	**703**	**22,259**
Males	**20,326**	**316**[2]	**15.5**[2]	**3,174**[2]	**156,174**[2]
Under 15 years	**223**	**15**	**67.0**	**126**	**566,048**
15-44 years	**1,992**	**82**	**41.1**	**1,508**	**757,009**
45-64 years	**4,911**	**102**	**20.9**	**1,254**	**225,241**
65 years & over	**13,200**	**117**	**8.8**	**287**	**21,725**
Females	**24,461**	**385**[2]	**15.8**[2]	**2,181**[2]	**89,176**[2]
Under 15 years	**217**	**16**	**73.5**	**95**	**438,981**
15-44 years	**1,593**	**75**	**47.2**	**841**	**527,836**
45-64 years	**4,265**	**107**	**25.1**	**829**	**194,341**
65 years & over	**18,386**	**187**	**10.2**	**416**	**22,642**

Note: *Numbers may not add to totals due to rounding.*
[1]Discounted at 4%.
[2]Excludes deaths with age not stated.

Table 6: Mortality from All Musculoskeletal Conditions: Number of Deaths, Persons Years Lost, and Discounted Productivity Losses by Age and Gender, 1988

Age	Number of deaths	Person years lost (thousands)	Productivity losses[1] (millions)
Both genders, total	**44,787**	**701**[2]	**$5,356**[2]
Under 1 year	144	11	61
1-4 years	85	6	39
5-9 years	84	6	45
10-14 years	127	8	76
15-19 years	268	16	186
20-24 years	382	21	279
25-29 years	502	25	370
30-34 years	668	30	469
35-39 years	787	31	509
40-44 years	978	35	535
45-49 years	1,236	38	540
50-54 years	1,686	45	546
55-59 years	2,524	57	540
60-64 years	3,731	70	457
65-69 years	5,156	79	330
70-74 years	6,065	74	199
75-79 years	6,527	62	108
80-84 years	5,981	42	49
85 years & over	7,852	46	18
Not stated	4		

Note: *Numbers may not add to totals due to rounding.*
[1]*Discounted at 4%.*
[2]*Excludes deaths with age not stated.*

Table 7: Mortality from All Musculoskeletal Conditions: Number of Deaths, Persons Years Lost, and Discounted Productivity Losses by Age: Males, 1988

Age	Number of deaths	Person years lost (thousands)	Productivity losses[1] (millions)
Males, total	**20,326**	**316[2]**	**$3,174[2]**
Under 1 year	68	5	33
1-4 years	48	3	25
5-9 years	50	3	29
10-14 years	58	4	40
15-19 years	145	8	113
20-24 years	199	10	167
25-29 years	275	13	233
30-34 years	369	15	302
35-39 years	460	17	345
40-44 years	545	18	349
45-49 years	677	19	346
50-54 years	923	23	344
55-59 years	1,384	28	325
60-64 years	1,926	32	239
65-69 years	2,571	35	149
70-74 years	2,894	31	79
75-79 years	2,847	23	38
80-84 years	2,350	14	16
85 years & over	2,534	13	5
Not stated	4		

Note: *Numbers may not add to totals due to rounding.*
[1]*Discounted at 4%.*
[2]*Excludes deaths with age not stated.*

Table 8: Mortality from All Musculoskeletal Conditions: Number of Deaths, Persons Years Lost, and Discounted Productivity Losses by Age: Females, 1988

Age	Number of deaths	Person years lost (thousands)	Productivity losses[1] (millions)
Females, total	**24,461**	**385[2]**	**$2,181[2]**
Under 1 year	76	6	28
1-4 years	37	3	15
5-9 years	35	3	16
10-14 years	69	5	36
15-19 years	124	8	73
20-24 years	183	11	113
25-29 years	227	12	137
30-34 years	299	14	168
35-39 years	326	14	164
40-44 years	434	17	186
45-49 years	558	19	194
50-54 years	762	22	202
55-59 years	1,140	28	214
60-64 years	1,804	38	218
65-69 years	2,585	44	181
70-74 years	3,172	43	120
75-79 years	3,679	38	70
80-84 years	3,631	28	33
85 years & over	5,319	34	12
Not stated	0		

Note: *Numbers may not add to totals due to rounding.*
[1]*Discounted at 4%.*
[2]*Excludes deaths with age not stated.*

Males account for 45% of the deaths, 45% of the person years lost, and 59% of the productivity losses (Figure 4). About 71% of the deaths due to musculoskeletal conditions occur among persons aged 65 years and older. Because of the short life expectancy and low earnings of this age group, person years lost amount to 44% of total person years lost and mortality costs are only 13% of the total mortality costs. By contrast, 8% of the musculoskeletal deaths are among persons aged 15 to 44, accounting for 22% of person years lost and 44% of total productivity losses (Figure 5).

Figure 4: All Musculoskeletal Mortality Losses by Gender, 1988

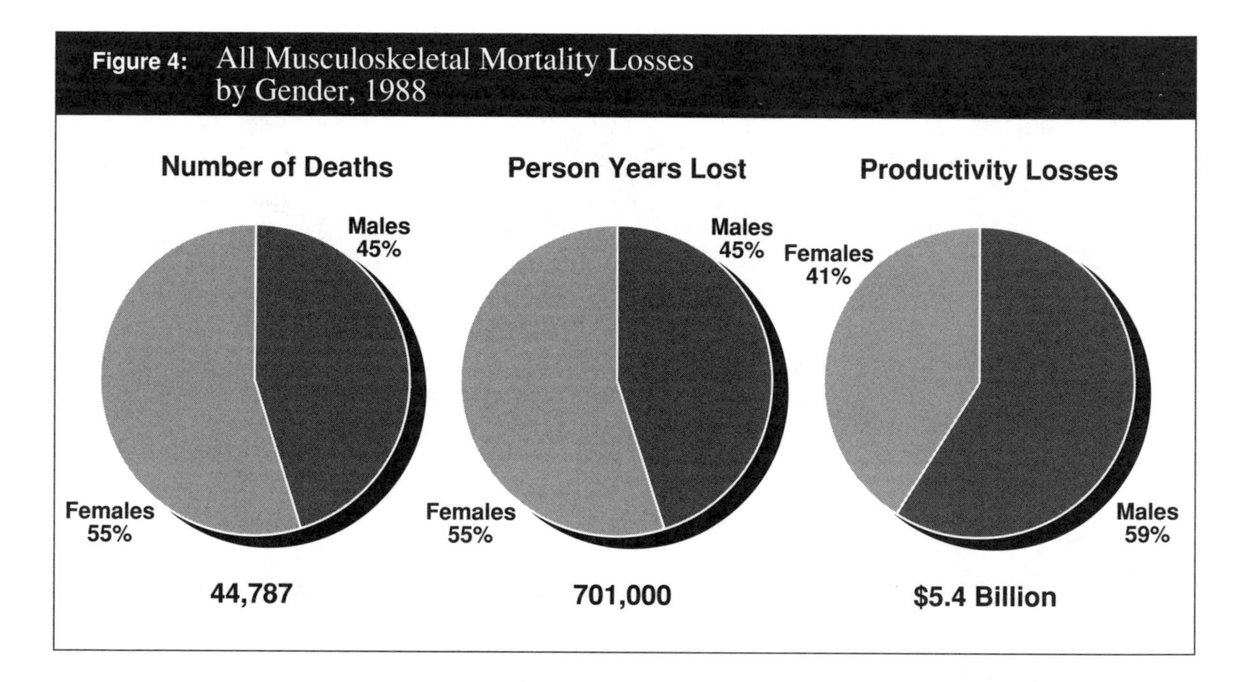

Figure 5: All Musculoskeletal Mortality Losses by Age, 1988

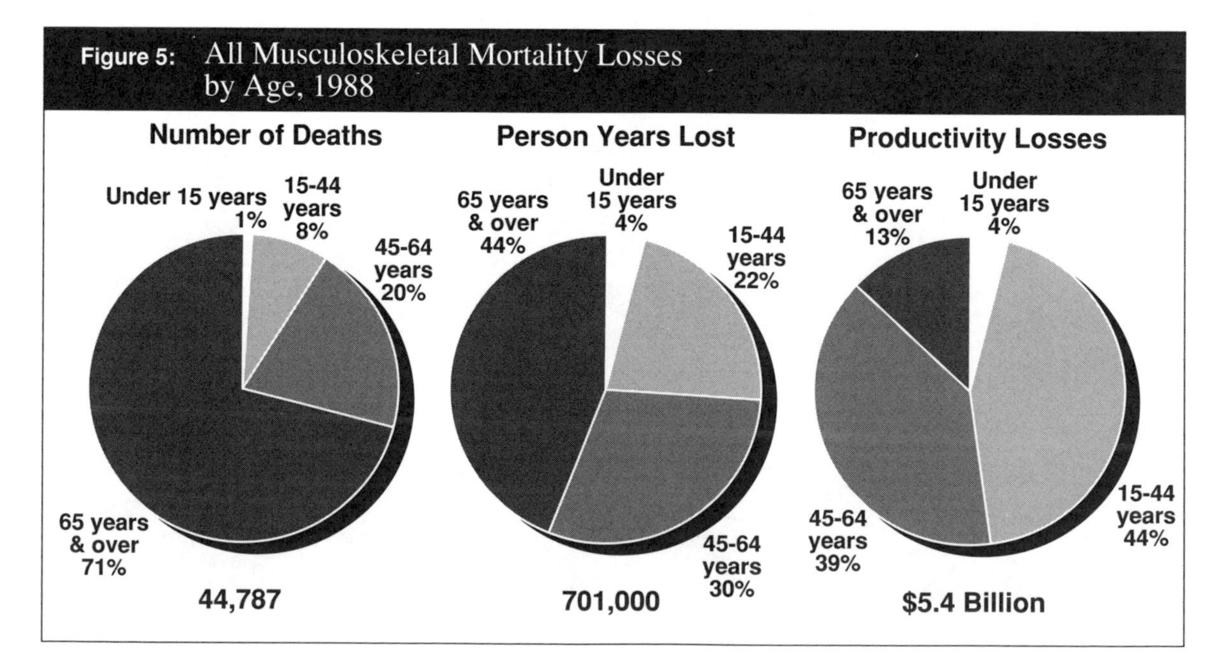

Musculoskeletal conditions less injuries

The estimated cost of musculoskeletal conditions less injuries is shown in Tables 9 through 11. These costs totaled almost $100 billion in 1988, representing 79% of the costs for all musculoskeletal conditions. The cost of musculoskeletal conditions less injuries for women is slightly higher than for men—$54.3 billion compared with $45.5 billion, respectively. The cost for the 45 to 64 year old age group comprises more than two-fifths of the total costs, followed by the cost for the 65 years old and older age group that comprise almost one-third of the total.

Table 9: Estimated Cost of All Musculoskeletal Conditions Less Injuries by Age and Type of Cost, 1988

(Cost in millions of dollars)

Type of cost	Total	Under 18 years	18-44 years	45-64 years	65 years & over
Total	**99,854**	**1,568**	**23,896**	**42,163**	**32,227**
Direct costs, total	**38,489**	**1,367**	**7,139**	**8,875**	**21,108**
Hospital inpatient	12,967	584	3,035	3,685	5,663
Hospital outpatient	3,280	147	768	932	1,433
Physician inpatient	1,850	107	518	529	696
Physician outpatient	2,307	177	763	731	636
Other practitioners	1,443	111	477	457	398
Drugs	451	5	70	177	199
Nursing home care	9,561	-	279	835	8,447
Prepayment/administration	1,851	66	343	427	1,015
Nonhealth sector	4,779	170	886	1,102	2,621
Indirect costs, total	**61,365**	**201**	**16,757**	**33,288**	**11,119**
Morbidity	56,616	-	14,720	31,396	10,500
Mortality[1]	4,749	201	2,037	1,892	619

[1] *Present value of lifetime earnings discounted at 4%.*

The distribution by type of cost shows that more than 60% of the total costs are indirect costs, mainly morbidity costs, and less than 40% are direct costs. Clearly, illness and disability caused by these conditions are high resulting in very high morbidity costs.

Table 10: Estimated Cost of All Musculoskeletal Conditions Less Injuries by Age and Type of Cost: Males, 1988

(Cost in millions of dollars)

Type of cost	Total	Under 18 years	18-44 years	45-64 years	65 years & over
Total	**45,519**	**917**	**12,357**	**21,512**	**10,733**
Direct costs, total	**14,964**	**804**	**3,642**	**3,776**	**6,742**
Hospital inpatient	5,675	357	1,667	1,652	1,999
Hospital outpatient	1,436	90	422	418	506
Physician inpatient	872	59	291	266	256
Physician outpatient	928	96	320	296	216
Other practitioners	580	60	200	185	135
Drugs	172	3	29	72	68
Nursing home care	2,723	-	86	236	2,401
Prepayment/administration	720	39	175	182	324
Nonhealth sector	1,858	100	452	469	837
Indirect costs, total	**30,555**	**113**	**8,715**	**17,736**	**3,991**
Morbidity	27,842	-	7,473	16,628	3,741
Mortality[1]	2,713	113	1,242	1,108	250

[1]*Present value of lifetime earnings discounted at 4%.*

Table 11: Estimated Cost of All Musculoskeletal Conditions Less Injuries by Age and Type of Cost: Females, 1988

(Cost in millions of dollars)

Type of cost	Total	Under 18 years	18-44 years	45-64 years	65 years & over
Total	**54,335**	**651**	**11,540**	**20,651**	**21,493**
Direct costs, total	**23,525**	**563**	**3,497**	**5,099**	**14,366**
Hospital inpatient	7,292	227	1,368	2,033	3,664
Hospital outpatient	1,844	57	346	514	927
Physician inpatient	978	48	227	263	440
Physician outpatient	1,379	81	443	435	420
Other practitioners	863	51	277	272	263
Drugs	279	2	41	105	131
Nursing home care	6,838	-	193	599	6,046
Prepayment/administration	1,131	27	168	245	691
Nonhealth sector	2,921	70	434	633	1,784
Indirect costs, total	**30,810**	**88**	**8,043**	**15,552**	**7,127**
Morbidity	28,774	-	7,247	14,768	6,759
Mortality[1]	2,036	88	796	784	368

[1]*Present value of lifetime earnings discounted at 4%.*

Musculoskeletal injuries

Musculoskeletal injuries include fractures, dislocations and sprains, open wounds and crushing injury, traumatic amputation, and other selected injuries affecting the musculoskeletal system. The cost of these injuries amounted to $26.1 billion in 1988 (Table 12). Women accounted for a greater percentage of these costs (53%) than did men (Tables 13 and 14). About 86% of the total are direct costs; the remaining 14% are indirect costs. Of the indirect costs totaling $3.6 billion, $3.0 billion (83%) are morbidity costs. The remaining $606 million are mortality costs. Mortality costs for musculoskeletal injuries are so low because only 8,346 deaths from injuries could be identified as musculoskeletal deaths. The National Center for Health Statistics only codes injury deaths according to the external causes of death, and the vast majority of injury deaths in 1988 (97,100) could not be identified according to body system. Thus, the mortality costs associated with musculoskeletal injuries reported here are clearly an understatement.

Table 12: Estimated Cost of Musculoskeletal Injuries by Age and Type of Cost, 1988

(Cost in millions of dollars)

Type of cost	Total	Under 18 years	18-44 years	45-64 years	65 years & over
Total	**26,107**	**2,010**	**8,567**	**4,318**	**11,212**
Direct costs, total	**22,498**	**1,990**	**6,155**	**3,412**	**10,941**
Hospital inpatient	9,170	754	2,533	1,389	4,494
Hospital outpatient	2,320	191	641	351	1,137
Physician inpatient	1,309	134	434	198	543
Physician outpatient	1,701	344	816	348	193
Other practitioners	1,064	215	510	218	121
Drugs	229	10	75	84	60
Nursing home care	2,830	-	86	236	2,508
Prepayment/administration	1,083	95	296	165	527
Nonhealth sector	2,792	247	764	423	1,358
Indirect costs, total	**3,609**	**20**	**2,412**	**906**	**271**
Morbidity[1]	3,003	-	2,101	716	186
Mortality[2]	606	20	311	190	85

[1]*Includes morbidity due to fractures, dislocations, sprains, and strains.*
[2]*Present value of lifetime earnings discounted at 4%.*

Table 13: Estimated Cost of Musculoskeletal Injuries by Age and Type of Cost: Males, 1988

(Cost in millions of dollars)

Type of cost	Total	Under 18 years	18-44 years	45-64 years	65 years & over
Total	12,178	1,328	5,895	1,991	2,964
Direct costs, total	9,889	1,315	4,183	1,508	2,883
Hospital inpatient	4,122	529	1,842	625	1,126
Hospital outpatient	1,043	134	466	158	285
Physician inpatient	650	87	320	100	143
Physician outpatient	904	205	474	161	64
Other practitioners	565	128	296	101	40
Drugs	109	6	44	39	20
Nursing home care	793	-	21	64	708
Prepayment/administration	476	63	201	73	139
Nonhealth sector	1,227	163	519	187	358
Indirect costs, total	2,289	13	1,712	483	81
Morbidity[1]	1,827	-	1,446	337	44
Mortality[2]	462	13	266	146	37

[1]Includes morbidity due to fractures, dislocations, sprains, and strains.
[2]Present value of lifetime earnings discounted at 4%.

Table 14: Estimated Cost of Musculoskeletal Injuries by Age and Type of Cost: Females, 1988

(Cost in millions of dollars)

Type of cost	Total	Under 18 years	18-44 years	45-64 years	65 years & over
Total	13,929	682	2,672	2,327	8,248
Direct costs, total	12,609	675	1,972	1,904	8,058
Hospital inpatient	5,048	225	691	764	3,368
Hospital outpatient	1,277	57	175	193	852
Physician inpatient	659	47	114	98	400
Physician outpatient	797	139	342	187	129
Other practitioners	499	87	214	117	81
Drugs	120	4	31	45	40
Nursing home care	2,037	-	65	172	1,800
Prepayment/administration	607	32	95	92	388
Nonhealth sector	1,565	84	245	236	1,000
Indirect costs, total	1,320	7	700	423	190
Morbidity[1]	1,176	-	655	379	142
Mortality[2]	144	7	45	44	48

[1]Includes morbidity due to fractures, dislocations, sprains, and strains.
[2]Present value of lifetime earnings discounted at 4%.

Costs of selected musculoskeletal conditions

The costs of selected musculoskeletal conditions, including arthritis, fractures, hip fractures, neoplasms, and congenital musculoskeletal deformities are shown in Table 15.

Table 15: Estimated Cost of Selected Musculoskeletal Conditions by Type of Cost, 1988

(Cost in millions of dollars)

Type of cost	Arthritis	Fractures	Hip Fractures	Neoplasms	Congenital Musculoskeletal Deformities
Total	**54,589**	**20,101**	**8,728**	**5,918**	**717**
Direct costs, total	**12,749**	**16,491**	**7,053**	**2,382**	**614**
Hospital inpatient	2,621	7,185	3,077	1,233	198
Hospital outpatient	663	1,818	778	312	52
Physician inpatient	349	997	385	168	35
Physician outpatient	581	463	18	37	29
Other practitioners	363	290	11	22	20
Drugs	145	66	5	7	4
Nursing home care	5,831	2,830	1,565	193	171
Prepayment/administration	613	794	339	116	28
Nonhealth sector	1,583	2,048	875	294	77
Indirect costs, total	**41,840**	**3,610**	**1,675**	**3,536**	**103**
Morbidity	[1]41,597	[2]3,003	1,415	na	na
Mortality[3]	243	607	260	3,536	103

[1]Includes morbidity for persons reporting arthritis and other musculoskeletal conditions.
[2]Includes morbidity due to fractures, dislocations, sprains, and strains.
[3]Present value of lifetime earnings discounted at 4%.
na- not available

Arthritis

Arthritis is the second most prevalent chronic condition (sinusitis is first) reported by respondents to the National Health Interview Survey. In 1988, 31.3 million conditions were reported, a rate of 130 per 1,000 persons. Arthritis increases with age, rising from 34 per 1,000 persons under 45 years of age, to 257 per 1,000 persons 45 to 64 years of age, and 486 per 1,000 persons 65 years of age and older (National Center for Health Statistics, 1989). Arthritis is the leading chronic condition reported by the elderly.

The high prevalence of arthritis in the population is reflected in high costs. In 1988, the total cost of arthritis is estimated at $54.6 billion, 43% of the total cost of all musculoskeletal conditions.

The debilitating and disabling effects of arthritis are seen in its high morbidity costs, amounting to $41.6 billion or 75% of total costs related to arthritis. Included in the morbidity costs are persons reporting arthritis and other chronic musculoskeletal conditions.

Direct costs of arthritis totaled $12.7 billion in 1988. Expenditures for nursing home care amounted to $5.8 billion, 46% of direct costs. Almost 20% of nursing home residents report rheumatoid arthritis, osteoarthritis and allied disorders, and other arthritis or rheumatism upon admission to the nursing home (National Center for Health Statistics, 1989). Hospital inpatient care totaled $2.6 billion, more than 20% of direct costs. The National Hospital Discharge Survey reports about 2.9 million days of care for patients hospitalized for arthritis and related disorders.

Fractures

The costs of fractures are estimated at $20 billion in 1988. Direct costs are more than 80% of the total costs. Of the total direct costs of $16.5 billion, hospital inpatient costs rank highest, $7.2 billion or 44%. Almost 900,000 persons were hospitalized for fractures, with an average length of stay of 8.8 days, or a total of 7.9 million days. The second highest category of direct costs is nursing home care, amounting to $2.8 billion, or 17% of the total. About 120,000 residents, 8% of total nursing home residents, were admitted to nursing homes in 1985 with fractures. Outpatient hospital care for fractures is the third highest direct cost, $1.8 billion, or 11% of the total.

Indirect costs are estimated at $3.6 billion, of which $3.0 billion are morbidity costs. Morbidity costs are based on bed disability days associated with acute conditions reported in the 1988 National Health Interview Survey and include fractures, dislocations, sprains, and strains. Therefore, morbidity costs for fractures may be slightly overstated. A total of 34.1 million bed days or 93,292 person years were reported for these conditions.

Hip fractures

The cost of hip fractures is estimated at $8.7 billion, 43% of the total costs of fractures. Direct costs comprise 80% of the total. Hospital inpatient care amounted to $3.1 billion, the largest portion of direct costs (44%). A total of 253,796 persons were hospitalized for hip fractures in 1988, of whom 85% were 65 years of age and older. Average length of stay for all inpatient hospital discharges was 13.4 days.

The cost of nursing home care is second highest amounting to $1.6 billion (22%). The National Nursing Home Survey reported 66,300 admissions in 1985 for hip fracture, 4.4% of total admissions.

Neoplasms

Included in this category are malignant and benign neoplasms of the musculoskeletal system, such as neoplasms of bone and articular cartilage, connective and other soft tissue, multiple myeloma and immunoproliferative neoplasms, and other malignant lymphomas. Total costs of neoplasms of the musculoskeletal system are estimated at $5.9 billion, excluding morbidity costs which could not be estimated because reliable data on morbidity were not available. 40% of the total costs are direct costs ($2.4 billion) and 60% ($3.5 billion) are mortality costs. Hospital inpatient care amounted to 1.4 million days at an estimated total cost of $1.2 billion.

A total of 26,783 deaths from neoplasms of the musculoskeletal system occurred in 1988. These deaths represent 440 million person years lost, or 16.4 years per death and a loss of $3.5 billion to the economy, or $132,041 per death.

Congenital musculoskeletal deformities

The total cost of musculoskeletal deformities is estimated at $717 million, excluding morbidity costs. Again, these costs could not be estimated because of the lack of reliable data on which to make estimates. Congenital musculoskeletal deformities clearly result in illness, disability, and reduced productivity. Thus, the total costs reported here are an underestimate. Direct costs are estimated at $6[illegible],000, of which hospital inpatient costs comprise 32% and nursing home costs 28%.

Conclusions

The total costs of musculoskeletal conditions to society in terms of resources used and in lost productivity are high, estimated at $126 billion in 1988. As noted earlier, these costs do not include costs associated with pain and suffering.

The total costs of musculoskeletal conditions for 1977, reported by Holbrook, et al, 1984, were $65.4 billion.. Direct costs were estimated at $53.7 billion and morbidity costs at $11.7 billion. Mortality costs were not estimated in that study. An update of costs from 1977 to 1988 would be expected to reflect an increase due to inflation as well as any change in the prevalence of musculoskeletal conditions. With no change in prevalence or in estimating methodology, direct expenditures for musculoskeletal conditions might be expected to increase about 222%, the rise in national personal health care expenditures during this period (Gibson, et al, 1984 and Lazenby, et al, 1990). In fact, the direct expenditures are only 43% higher. Morbidity costs in the current study, on the other hand, are significantly higher than those of the previous study—410%.

The main sources of difference between the two sets of estimates for most of the cost categories are the use in the current study of new and current national surveys as well as the use of different methodologies. For example, morbidity costs in the earlier study were based on bed disability days attributable to musculoskeletal conditions. In the current study, bed disability days were used only for acute musculoskeletal conditions. For chronic conditions, however, prevalence of musculoskeletal conditions was used, adjusted for inability to work or perform major activity. Thus, the two sets of estimates are not comparable.

References

American Hospital Association. Hospital Statistics, 1989-90 Edition. Chicago, IL: American Hospital Association, 1989.

American Hospital Association. Selected Community Hospital Statistics: 1981-90, *Health Care Financing Review*, 1991; 12:146.

American Medical Association Center for Health Policy Research, Physician Marketplace Statistics, 1989: Profiles for Detailed Specialties, Selected States and Practice Arrangements. Chicago, IL, 1989;p48.

Douglass JB, Kenney GM, Miller TR. Which estimates of household production are best? *Journal of Forensic Economics*, 1990; 4:25-45.

Gibson HC, Levit KR, Lazenby HC, Waldo DR. National health expenditures, 1983. *Health Care Financing Review*, 1984; 6(1):1-29.

Hodgson TA, Meiners M. Cost-of-illness methodology: A guide to current practices and procedures. *Milbank Memorial Fund Quarterly*, 1982; 60:429-462.

Holbrook TL, Grazier K, Kelsey JL, et al. *The Frequency of Occurrence, Impact and Cost of Selected Musculoskeletal Conditions in the United States.* American Academy of Orthopaedic Surgeons, Park Ridge, IL, 1984.

LaPlante MP. Disability risks of chronic illnesses and impairments. Disability Statistics Report. Report 2. National Institute on Disability and Rehabilitation Research, U.S. Department of Education, June 1991.

Latta VB and Helbing C. Short-stay hospital service by diagnosis-related groups. *Health Care Financing Review*, 1991; 12:105-140.

Lazenby HC, Letsch SW. National health expenditures, 1989. *Health Care Financing Review*, 1990; 12(2):1-26.

Murt HA, Parsons PE, Harlan WR, et al: Disability utilization and costs associated with musculoskeletal conditions, United States, 1980. National Medical Care Utilization and Expenditure Survey. Series C, Analytical Report No. 5, DHHS Pub. No. 86-20405. National Center for Health Statistics, Public Health Service. Washington. U.S. Government Printing Office, Sept. 1986.

Mushkin SJ, Landefeld S. Non-health sector costs of illness. *Report A7.* Washington, DC: Public Services Laboratory, Georgetown University, 1978.

National Center for Health Statistics. Current estimates from the National health interview survey, 1988. *Vital and Health Statistics.* Series 10: No. 173, DHHS Pub. No. (PHS)89-1501. Public Health Service, Hyattsville, MD, October 1989.

National Center for Health Statistics. The National Nursing Home Survey; 1985 summary for the United States. *Vital and Health Statistics.* DHHS Pub. No. (PHS)89-1758. Washington, U.S. Government Printing Office, 1989.

National Center for Health Statistics. Unpublished data from the National Hospital Discharge Survey, 1988. U.S. Department of Health and Human Services, Hyattsville, MD.

National Center for Health Statistics. Unpublished data from public use tapes, National Ambulatory Care Survey, 1985. U.S. Department of Health and Human Services, Hyattsville, MD.

National Center for Health Statistics. Unpublished data from public use tapes, National Nursing Home Survey, 1985. U.S. Department of Health and Human Services, Hyattsville, MD.

Rice DP, Hodgson TA, Kopstein AN. The economic costs of illness: A replication and update. *Health Care Financing Review* 1985; 7:61-80

U.S. Bureau of the Census. Money income of household, families, and persons in the United States, 1985. *Current Population Reports*, Series P-25, No. 1000. Washington, DC: US Government Printing Office, 1987.

Yelin EH, Katz PP. Transitions in health status among community-dwelling elderly people with arthritis. *Arthritis and Rheumatism*, 1990; 33:pp1205-1215.

Yelin EH, Katz PP. Labor force participation among persons with musculo-skeletal conditions, 1970-1987. *Arthritis and Rheumatism*, 1991; 34: pp1361-1370.

Appendices

Appendix A

Table 1

First-listed Diagnosis by ICD•9•CM Code
1988 National Hospital Discharge Survey
Musculoskeletal Conditions (includes Injuries)

		Hospitalizations Age Category				
		Total	Less than 18	18-44 years	45-64 years	65 & over
015	Tuberculosis of Bones and Joints	*	*	*		*
098+	Gonococcal Infections	*	*	*		
170+	Malignant Neoplasm of Bone and Articular Cartilage	13,000	7,000	2,000	*	*
171+	Malignant Neoplasm of Connective and Other Soft Tissue	9,000	*	3,000	*	5,000
198+	Secondary Malignant Neoplasm	61,000	*	3,000	23,000	34,000
200+	Other Named Variants of Lymphosarcoma	3,000	*	*	*	2,000
202+	Other Malignant Lymphomas	14,000	*	*	6,0000	6,000
203	Multiple Myeloma and Immunoproliferative Neoplasms	19,000	*	*	5,000	13,000
213+	Benign Neoplasm	6,000	*	2,000	*	*
214+	Lipoma	9,000	*	5,000	3,000	*
215+	Other Benign Neoplasm of Connective and Other Soft Tissue	4,000	*	*	*	*
238+	Neoplasm of Uncertain Behavior	2,000	*	*	*	*
239+	Neoplasm of Unspecified Nature	*	*	*		*
252	Disorders of Parathyroid Gland	5,000		*	2,000	2,000
268	Vitamin D Deficiency	*	*	*		*
274	Gout	10,000		*	3,000	7,000
354	Mononeuritis of Upper Limb and Mononeuritis Multiplex	32,000	*	10,000	15,000	6,000

356	Hereditary and Ideopathic Peripheral Neuropathy	3,000		*	*	*
357	Inflammatory and Toxic Neuropathy	9,000	2,000	*	3,000	2,000
443+	Other Peripheral Vascular Disease	*		*	*	
446	Polyarteritis Nodosa and Allied Conditions	9,000	3,000	*	*	3,000
696+	Psoriasis and Similar Disorders	*		*	*	*
710	Diffuse Diseases of Connective Tissue	21,000	*	8,000	7,000	4,000
711	Arthropathy Associated with Infections	23,000	4,000	7,000	7,000	5,000
714	Rheumatoid Arthritis and Other Inflammatory Polyarthropathies	48,000	*	5,000	21,000	20,000
715	Osteoarthrosis and Allied Disorders	205,000	*	8,000	50,000	146,000
716	Other and Unspecified Arthropathies	24,000	*	7,000	9,000	7,000
717	Internal Derangement of the Knee	52,000	5,000	36,000	9,000	*
718	Other Derangement of Joint	59,000	13,000	38,000	6,000	2,000
719	Other and Unspecified Disorders of Joint	27,000	5,000	7,000	6,000	9,000
720	Ankylosing Spondylitis and Other Inflammatory Spondylopathies	8,000	*	4,000	3,000	*
721	Spondylosis and Allied Disorders	75,000	*	14,000	26,000	35,000
722	Intervertebral Disk Disorders	417,000	2,000	222,000	142,000	51,000
723	Other Disorders of Cervical Region	35,000	*	17,000	13,000	4,000
724	Other and Unspecified Disorders of Back	178,000	*	69,000	63,000	45,000
725	Polymyalgia Rheumatica	4,000		*	*	3,000
726	Peripheral Enthesopathies and Allied Syndromes	48,000	*	23,000	18,000	6,000
727	Other Disorders of Synovium, Tendon and Bursa	59,000	7,000	20,000	22,000	8,000
728	Disorders of Muscle, Ligament and Fascia	25,000	3,000	7,000	8,000	6,000
729	Other Disorders of Soft Tissues	55,000	*	22,000	17,000	14,000
730	Osteomyelitis, Periostitis and Other Infections Involving Bone	39,000	5,000	8,000	13,000	13,000
731	Osteitis Deformans and Osteopathies Associated with Other Disorders	*	.	*	*	
732	Osteochondropathies	13,000	9,000	3,000	*	
733	Other Disorders of Bone and Cartilage	152,000	7,000	36,000	31,000	77,000
734	Flat Foot	*		*	*	*
735	Acquired Deformities of Toe	39,000	*	12,000	15,000	11,000
736	Other Acquired Deformities of Limbs	16,000	5,000	5,000	4,000	*

737	Curvature of Spine	7,000	4,000	3,000	*	*
738	Other Acquired Musculoskeletal Deformity	17,000	*	10,000	4,000	2,000
739	Nonallopathic Lesions of Head Region, not elsewhere classfied	*	*	*		
741	Spina Bifida	2,000	*	*	*	
754+	Certain Congenital Musculoskeletal Deformities	17,000	15,000	*	*	*
755	Other Congenital Anomalies of Limbs	11,000	8,000	*	*	*
756+	Other Congenital Musculoskeletal Anomalies	11,000	3,000	3,000	3,000	2,000
781	Symptoms Involving Nervous and Musculoskeletal Systems	4,000	*	*	*	*
805	Fracture of Vertebral Column w/o Mention of Spinal Cord	76,000	6,000	25,000	13,000	32,000
806	Fracture of Vertebral Column with Spinal Cord Injury	7,000	*	4,000	*	*
807	Fracture of Rib(s), Sternum, Larynx and Trachea	42,000	*	12,000	10,000	19,000
808	Fracture of Pelvis	51,000	3,000	15,000	5,000	27,000
809	Ill-defined Fracture of Bones of Trunk	*				*
810	Fracture of Clavicle	7,000	*	5,000	*	*
811	Fracture of Scapula	2,000	*	*	*	*
812	Fracture of Humerus	62,000	24,000	10,000	7,000	21,000
813	Fracture of Radius and Ulna	81,000	21,000	27,000	17,000	16,000
814	Fracture of Carpal Bone(s)	7,000	6,000	*	*	
815	Fracture of Metacarpal Bone(s)	8,000	*	5,000	*	*
816	Fracture of one or more Phalanges of Hand	17,000	3,000	11,000	3,000	*
817	Multiple Fractures of Hand Bones	*	*	*	*	*
818	Ill-Defined Fractures of Upper Limb	*	.		*	
820	Fracture of Neck of Femur	254,000	4,000	9,000	24,000	217,000
821	Fracture of Other and Unspecified Parts of Femur	53,000	20,000	12,000	6,000	15,000
822	Fracture of Patella	23,000	2,000	9,000	5,000	7,000
823	Fracture of Tibia and Fibula	76,000	13,000	36,000	13,000	14,000
824	Fracture of Ankle	102,000	16,000	46,000	28,000	12,000
825	Fracture of One or More Tarsal and Metatarsal Bones	24,000	3,000	14,000	5,000	3,000

826	Fracture of One or More Phalanges of Foot	4 ,000	*	*	*	*
827	Other, Multiple and Ill-Defined Fractures of Lower Limb	*	.	*	.	.
828	Multiple Fractures Involving Both Lower Limbs Lower w/Upper, etc	*	.	*	.	*
829	Fracture of Unspecified Bones	*	.	.	*	.
831	Dislocation of Shoulder	10,000	*	6,000	2,000	*
832	Dislocation of Elbow	3,000	*	*	*	*
833	Dislocation of Wrist	*	*	*	*	*
834	Dislocation of Finger	4,000	*	2,000	*	*
835	Dislocation of Hip	3,000	*	*	*	*
836	Dislocation of Knee	35,000	4,000	22,000	7,000	2,000
837	Dislocation of Ankle	*	*	*	*	*
838	Dislocation of Foot	*	*	*	*	*
839	Other, Multiple, Ill-Defined Dislocations	7,000	*	3,000	*	*
840	Sprains and Strains of Shoulder and Upper Arm	32,000	*	7,000	18,000	8,000
841	Sprains and Strains of Elbow and Forearm	*	*	*	*	.
842	Sprains and Strains of Wrist and Hand	5,000	*	3,000	*	*
843	Sprains and Strains of Hip and Thigh	5,000	*	*	*	*
844	Sprains and Strains of Knee and Leg	38,000	6,000	22,000	8,000	2,000
845	Sprains and Strains of Ankle and Foot	10,000	*	7,000	*	*
846	Sprains and Strains of Sacroiliac Region	42,000	*	24,000	10,000	7,000
847	Sprains and Strains of Other and Unspecified Parts of Back	55,000	*	37,000	12,000	5,000
848+	Other and Ill-Defined Sprains and Strains	4,000	.	3,000	*	*
874	Open Wound of Neck	8,000	*	6,000	*	*
875	Open Wound of Chest Wall	12,000	*	10,000	*	*
876	Open Wound of Back	2,000	*	*	*	*
877	Open Wound of Buttock	*	*	*	.	.
879+	Open Wound of Other and Unspecified Sites, Except Limbs	13,000	2,000	10,000	*	*
880	Open Wound of Shoulder and Upper Arm	4,000	*	3,000	*	.
881	Open Wound of Elbow, Forearm and Wrist	16,000	*	13,000	*	*
882	Open Wound of Hand Except Finger(s) Alone, w/o Mention of Complication	21,000	4,000	15,000	*	*
883	Open Wound of Finger(s)	23,000	4,000	14,000	4,000	*

884	Multiple and Unspecified Open Wound of Upper Limb	4,000	*	3,000	*	*
885	Traumatic Amputation of Thumb (Complete) (Partial)	2,000	*	2,000	*	*
886	Traumatic Amputation of Other Finger(s) (Complete) (Part	13,000	*	9,000	2,000	*
887	Traumatic Amputation of Arm and Hand (Complete) (Partial)	*	*	*	.	*
890	Open Wound of Hip and Thigh	8,000	*	5,000	*	*
891	Open Wound of Knee, Leg (Except Thigh), and Ankle	18,000	4,000	10,000	2,000	*
892	Open Wound of Foot Except Toe(s) Alone	13,000	6,000	3,000	*	*
893	Open Wound of Toe(s)	2,000	*	*	.	*
894	Multiple and Unspecified Open Wound of Lower Limb	*	*	*	*	*
895	Traumatic Amputation of Toe(s) (Complete) (Partial)	2,000	*	*	*	*
897	Traumatic Amputation of Leg(s) (Complete) (Partial)	*	*	*	*	*
905+	Late Effects of Musculoskeletal and Connective Tissue Injuries	6,000	*	3,000	*	*
922	Contusion of Trunk	24,000	4,000	11,000	4,000	5,000
923	Contusion of Upper Limb	*	*	*	*	*
924	Contusion of Lower Limb and of Other and Unspecified Sites	43,000	6,000	20,000	5,000	12,000
926	Crushing Injury of Trunk	*	*	*		
927	Crushing Injury of Upper Limb	5,000	*	3,000	*	.
928	Crushing Injury of Lower Limb	4,000	*	3,000	*	*
955	Injury to Peripheral Nerve(s) of Shoulder Girdle and Upper Limb	5,000	*	3,000	*	*
956	Injury to Peripheral Nerve(s) of Pelvic Girdle and Lower Limb	*	.	*	*	.
957+	Injury to Other and Unspecified Nerves	*	.	*	.	.
959+	Injury, Other and Unspecified	19,000	4,000	9,000	2,000	4,000
996+	Complications Peculiar to Certain Specified Procedures	172,000	9,000	37,000	49,000	76,000
997+	Complications Affecting Specified Body Systems Not Elsewhere Classified	8,000	*	*	3,000	5,000

V10	Fracture of Carpal Bone(s)	*			*	
V13	Personal History of Other Diseases	*			*	
V52	Fitting and Adjustment of Prosthetic Device	*			*	
V67	Follow-up Examination	*				*
Total		3,519,000	321,000	1,229,000	863,000	1,105,000

+ Some elements within this code were not included.

* Estimate does not meet standards of reliability or precision.

Table 2

First-listed Diagnosis by ICD•9•CM Code
1988 National Hospital Discharge Survey

Musculoskeletal Conditions (includes Injuries)
30 Most Frequent Conditions by ICD•9•CM Code (3 Digit)

Sorted by frequency:		Hospitalizations
722	Intervertebral Disk Disorders	417,000
820	Fracture of Neck of Femur	254,000
715	Osteoarthrosis and Allied Disorders	205,000
724	Other and Unspecified Disorders of Back	178,000
996*	Complications Peculiar to Certain Specified Procedures	172,000
733	Other Disorders of Bone and Cartilage	152,000
824	Fracture of Ankle	102,000
813	Fracture of Radius and Ulna	81,000
805	Fracture of Vertebral Column w/o Mention of Spinal Cord Injury	76,000
823	Fracture of Tibia and Fibula	76,000
721	Spondylosis and Allied Disorders.	75,000
812	Fracture of Humerus	62,000
198*	Secondary Malignant Neoplasm	61,000
718	Other Derangement of Joint	59,000
727	Other Disorders of Synovium, Tendon and Bursa	59,000
729	Other Disorders of Soft Tissues	55,000
847	Sprains and Strains of Other and Unspecified Parts of Back	55,000
821	Fracture of Other and Unspecified Parts of Femur	53,000
717	Internal Derangement of the Knee	52,000
808	Fracture of Pelvis	51,000

726	Peripheral Enthesopathies and Allied Syndromes	48,000
714	Rheumatoid Arthritis and Other Inflammatory Polyarthropathies	48,000
924	Contusion of Lower Limb and of Other and Unspecified Sites	43,000
807	Fracture of Rib(s), Sternum, Larynx and Trachea	42,000
846	Sprains and Strains of Sacroiliac Region	42,000
735	Acquired Deformities of Toe	39,000
730	Osteomyelitis, Periostitis and Other Infections Involving Bone	39,000
844	Sprains and Strains of Knee and Leg	38,000
836	Dislocation of Knee	35,000
723	Other Disorders of Cervical Region	35,000

* Some elements within this code were not included.

Table 3

All Listed Procedures by ICD•9•CM Code
1988 National Hospital Discharge Survey

Operations on the Musculoskeletal System (excluding Jaw), 7700-8499
30 Most Frequent Operations by ICD•9•CM Procedure Code

Sorted by frequency ICD-•9•CM Procedure Code		Procedures
8051	Excision of Intervertebral Disk	242,000
7935	Open Reduction of Fracture of Femur with Internal Fixation	159,000
8026	Arthroscopy of Knee	123,000
7936	Open Reduction of Fracture of Tibia and Fibula with Internal Fixation	114,000
8141	Total Knee Replacement	105,000
8159	Other Total Hip Replacement (wlthout use of Methyl Methacrylate)	81,000
806	Excision of Semilunar Cartilage of Knee	62,000
8191	Arthrocentesis	49,000
8151	Total Hip Replacement with use of Methyl Methacrylate	48,000
7902	Closed Reduction of Fracture of Radius and Ulna without Internal Fixation	45,000
8162	Other Replacement of Head of Femur (without use of Methyl Methacrylate)	45,000
8411	Amputation of Toe	43,000
7906	Closed Reduction of Fracture of Tibia and Fibula	43,000
8147	Other Repair of Knee	43,000
8145	Other Repair of the Cruciate Ligaments	42,000
8102	Other Cervical Spinal Fusion	39,000
8183	Other Repair of Shoulder	38,000
8363	Rotator Cuff Repair	37,000

7779	Excision of Other Bone for Graft, except Facial Bones	36,000
7865	Removal of Internal Fixation Device from Femur	35,000
7932	Open Reduction of Fracture of Radius and Ulna without Internal Fixation	34,000
8417	Amputation above Knee	32,000
7931	Open Reduction of Fracture of Humerus with Internal Fixation	30,000
7855	Internal Fixation of Femur without Fracture Reduction	29,000
8415	Other Amputation below Knee	25,000
7759	Other Bunionectomy	25,000
8086	Other Local Excision or Destruction of Lesion of Knee	23,000
7915	Closed Reduction of Fracture of Femur with Internal Fixation	21,000
7939	Open Reduction of Fracture of Other Specified Bone, except Facial Bones, with Internal Fixation	21,000
7789	Other Partial Ostectomy of Other Bones, except Facial Bones	20,000

Appendix B

Table 1

ICD•9•CM Diagnosis Codes for Musculoskeletal Conditions by Aggregate Category

Musculoskeletal Diseases and Connective Tissue Disorders	710-739
Injuries:	
Fractures (excluding skull)	805-829
Dislocations and Sprains (excluding jaw)	831-847, 848.3-848.9
Crushing Injury	926-928
Traumatic Amputation	885-887, 895-897
Open Wound	874-877, 879.2-884, 890-894
Contusion	922-924
Other Injury	954-956, 957.1-957.9, 959.1-959.9
Complications or Late Effects:	
Mechanical Complication of Internal Orthopaedic Device, Implant and Graft	996.4
Infection and Inflammatory Reaction due to Internal Prosthetic Device, Implant and Graft	996.6-996.7
Complications of Reattached Extremity or Body Part	996.9
Late Amputation Stump Complication	997.6
Late Effects of Musculoskeletal and Connective Tissue Injuries	905.1-905.9
Congenital Musculoskeletal Deformities and Anomalies	754.1-755, 756.1-756.9, 741
Neoplasms:	
A. Malignant	170.2-170.9, 171.2-171.9, 198.5, 200.82-200.85,202.82-202.85, 203

B. Benign	213.2-213.9, 215.2-215.9, 238.0-238.1, 239.2
Other	015, 098.5, 252.0-252.1, 268.1-268.2, 274, 354, 356.9, 357.0, 443.0, 446, 696.0, 781.0, 781.2-781.9
V Codes:	V10.81, V13.5, V49, V52.1, V53.7, V57.81, V66.4, V67.4

ICD•9•CM Procedure Codes for Musculoskeletal Conditions

Operations on the Musculoskeletal System (excluding operations on facial bones and joints)	77.0-84.46, 03.02, 03.53

Table 2

ICD•9•CM Diagnosis Code for Specific Musculoskeletal Conditions

Back Pain (excluding cervical)

Back Disorders	
720*	Ankylosing spondylitis and other inflammatory spondylopathies
721.2-.9	Spondylosis and allied disorders
724	Other and unspecified disorders of back
Disk Disorders	
722.1-.2	Displacement of intervertebral disk
722.3	Schmorl's nodes
722.5-.6	Degeneration of intervertebral disk
722.72-.79	Intervertebral disk disorder with myelopathy
722.80, .82-.83	Postlaminectomy syndrome
722.90, .92-.93	Other and unspecified disk disorder
Back Injury	
805.2-.8	Closed fracture of vertebra without mention of spinal cord injury
806.2-.8	Closed fracture of vertebra with spinal cord injury
839.2-.4	Closed dislocation, vertebra
846	Sprains and strains of sacroiliac region
847.1-.9	Other sprains and strains of back

Neck Pain

Neck disorders	
721.0-.1	Cervical spondylosis
723	Disorders of cervical region
Disk disorders	
722.0	Displacement of cervical intervertebral disk
722.4	Degeneration of cervical intervertebral disk
722.71	Intervertebral disk disorder, with myelopathy
722.81	Postlaminectomy syndrome of cervical region
722.91	Other and unspecified disk disorders of cervical region

Neck injury

805.0	Closed fracture of cervical vertebra without mention of spinal cord injury
806.0	Closed fracture of cervical vertebra with spinal cord injury
839.0	Closed dislocation, cervical vertebra
847.0	Neck sprain

Arthritis

098.50	Gonococcal arthritis
274.0	Gouty arthropathy
711	Arthropathy associated with infections
712	Crystal arthropathies
713	Arthropathy associated with other disorders classified elsewhere
714	Rheumatoid arthritis and other inflammatory polyarthropathies
715	Osteoarthrosis and allied disorders
716	Other and unspecified arthropathies
720*	Ankylosing spondylitis and other inflammatory spondylopathies

Congenital Musculoskeletal Deformities and Anomalies

741	Spina bifida
754.1-.89	Certain congenital musculoskeletal deformities
755	Other congenital anomalies of limbs
756	Other congenital musculoskeletal anomalies

Neoplasms

Malignant

170.2-.9	Malignant neoplasm of bone and articular cartilage (excluding skull, face and mandible)
171.2-.9	Malignant neoplasms of connective and other soft tissue
198.5	Secondary malignant neoplasm of bone and bone marrow
200.82-.85	Other named variants of lymphosarcoma and reticulosarcoma
202.82-.85	Other malignant lymphomas
203	Multiple myeloma and immunoproliferative neoplasms

*Some overlap between Back Pain and Arthritis codes

Benign

213.2-.9	Benign neoplasm of bone and articular cartilage
215.2-.9	Benign neoplasm of connective and other soft tissue
238.0	Neoplasm of uncertain behavior of bone and articular cartilage
238.1	Neoplasm of uncertain behavior of connective and other soft tissue
239.2	Neoplasm of unspecified nature of bone, soft tissue and skin

Musculoskeletal Injuries not Indicated Above

807	Fracture of rib(s), sternum, larynx and trachea
808	Fracture of pelvis
810	Fracture of clavicle
811	Fracture of scapula
812	Fracture of humerus
813	Radius and ulna
814	Fracture of carpal bone(s)
815	Fracture of metacarpal bone(s)
816	Fracture of one or more phalanges of hand
817	Multiple fracture of hand bones
818	Ill-defined fracture of upper limb
819	Multiple fracture involving both upper limbs and upper limb with rib(s) and sternum
820	Fracture of neck of femur
821	Fracture of other and unspecified parts of femur
822	Fracture of patella
823	Fracture of tibia and fibula
824	Fracture of ankle
825	Fracture of one or more tarsal and metatarsal bones
826	Fracture of one or more phalanges of foot
827	Other multiple and ill-defined Fracture of of lower limb
828	Other multiple and ill-defined fracture involving both lower limbs, lower with upper limb and lower limb(s) with rib(s) and sternum
829	Fracture of unspecified bones
831	Dislocation of shoulder
832	Dislocation of elbow
833	Dislocation of wrist
834	Dislocation of finger

835	Dislocation of hip
836	Dislocation of knee
837	Dislocation of ankle
838	Dislocation of foot
839.6-.9	Other multiple and ill-defined dislocations not otherwise listed
840	Sprains and strains of shoulder and upper arm
841	Sprains and strains of elbow and forearm
842	Sprains and strains of wrist and hand
843	Sprains and strains of hip and thigh
844	Sprains and strains of knee and leg
845	Sprains and strains of ankle and foot
847.0	Sprains and strains of other and unspecified parts of back
848.3-.9	Other and ill-defined musculoskeletal sprains and strains
874	Open wound of neck
875	Open wound of chest wall
876	Open wound of back
877	Open wound of buttock
879.2-.9	Open wound of other and unspecified sites, except limbs
880	Open wound of shoulder and upper arm
881	Open wound of elbow, forearm and wrist
882	Open wound of hand, except finger(s) alone
883	Open wound of finger(s)
884	Multiple and unspecified open wound of upper limb
885	Traumatic amputation of thumb (complete) (partial)
886	Traumatic amputation of other fingers (complete) (partial)
887	Traumatic amputation of arm and hand (complete) (partial)
890	Open wound of hip and thigh
891	Open wound of knee, leg (except thigh), and ankle
892	Open wound of foot except toe(s) alone
893	Open wound of toes
894	Multiple and unspecified open wound of lower limb
895	Traumatic amputation of toe(s) (complete) (partial)
896	Traumatic amputation of foot (complete) (partial)
897	Traumatic amputation of leg(s) (complete) (partial)
922	Contusion of trunk

923	Contusion of upper limb
924	Contusion of lower limb and of other unspecified sites
926	Crushing injury of trunk
927	Crushing injury of upper limb
928	Crushing injury of lower limb
954	Injury to other nerve(s) of trunk excluding shoulder and pelvic girdle
955	Injury to peripheral nerve(s) of shoulder girdle and upper limb
956	Injury to peripheral nerve(s) of pelvic girdle and lower limb
957.1-.9	Injury to other and unspecified nerves
959.1-.9	Injury, other and unspecified
015	Tuberculosis of bones and joints
252.0	Hyperparathyroidism
252.1	Hypoparathyroidism
268.1	Rickets, late effect
268.2	Osteomalacia, unspecified
274.1-.9	Gout
354	Mononeuritis of upper limb and mononeuritis multiplex
356.9	Unspecified idiopathic peripheral neuropathy
357.0	Acute infective polyneuritis
443.0	Raynauds syndrome
446	Polyarteritis nodosa and allied conditions
696.0	Psoriatic arthropathy
781.0, .2-.9	Symptoms involving nervous and musculoskeletal systems
905.1-.9	Late effects of musculoskeletal and connective tissue injury
996.4	Mechanical complication of internal orthopaedic device, implant and graft
996.6-.7	Infection and inflammatory reaction to internal prosthetic device, implant and graft
996.9	Complications of reattached extremity or body part
997.6	Late amputation stump complication
V10.81	Personal history of malignant neoplasm of bone
V13.5	Personal history of other musculoskeletal disorders
V49	Problems with limbs and other problems
V52.1	Fitting and adjustment of artificial leg (complete) (partial)
V53.7	Fitting and adjustment of orthopaedic devices
V57.81	Care involving orthotic training
V66.4	Convalescence following treatment of fracture
V67.4	Follow-up examination following treatment of fracture

Data Sources

The major sources of data used in this publication were four surveys conducted by the National Center for Health Statistics (NCHS). These surveys were (1) the National Health Interview Survey (NHIS), which is a continuing nationwide survey of the noninstitutional population conducted by household interview, (2) the National Hospital Discharge Survey (NHDS), which samples records of patients discharged from short-stay general hospitals, (3) the National Ambulatory Medical Care Survey (NAMCS), a continuing periodic survey in which records of visits to office-based physicians are sampled, and (4) the National Nursing Home Survey (NNHS), a continuing periodic survey of nursing homes.

These surveys are based on a scientifically-selected probability sample of the United States noninstitutional population, nursing home residents, or on samples of the records of short-stay hospitals or physicians. These surveys have the desirable characteristics of being directly representative of their respective elements of the United States population or of hospitals and physicians within the United States. In addition, while there have been some changes in the data elements collected over time, these surveys have collected a large, relatively consistent volume of data since at least the early 1970s. When these data sets are used to complement each other, they provide a comprehensive profile of the frequency of musculoskeletal conditions and their associated impact and health care utilization.

A major objective of this publication has been to include a greater variety of data sources. The NCHS is limited in the number of topics and level of detail that can be addressed in the four surveys listed above. Data from other NCHS sources such as the Supplements on Aging and Medical Device Implants appended to the NHIS as well as other NCHS surveys such as the second National Health and Nutrition Examination Survey (NHANESII) and the NHANESI Epidemiologic Follow-up Study have also been used.

More extensive use has also been made of the results of studies reported in scientific and epidemiologic journals as secondary sources of data.

Other national databases including the Health Care Financing Administration's Medicare Procedure File and Consolidated Consulting Group's Medicare Hospital Utilization Data Bases have also been used.

It is important to recognize that any one source of data provides an incomplete picture of the frequency and impact of a condition or disease. Interview surveys, for instance, generally underestimate the frequency of most diseases. However, information obtained through the use of interviews is essential in order to assess an individual's symptoms, or how that individual feels he is affected by a disease. In addition, for some conditions, this is the only possible means of diagnosis.

In contrast, the use of objective methods, such as physical examinations, laboratory measurements, and x-rays, for the detection and diagnosis of disease is not dependent on the subjective reporting of symptoms. However, the objective evidence used to establish the presence or absence of disease does not always correlate with reported symptoms. It is clear that numerous individuals who report symptoms do have evidence of disease upon examination, and conversely some individuals who have objective evidence of disease do not experience symptoms. Thus, although objective measures are valuable tools, they may also yield an incomplete picture of the impact of a given disorder.

Data obtained from medical records are also subject to certain limitations. Many persons affected with certain musculoskeletal disorders do not seek medical care. Although records of visits to physicians and hospitals may provide estimates of the volume of visits, these records exclude those individuals who do not seek medical care. Therefore, data which are based on existing records will yield only a partial understanding of the ways various musculoskeletal disorders affect the population.

Medical records may also not indicate the underlying musculoskeletal condition. For instance, fractures at various sites, especially hip fractures, are a major consequence of osteoporosis. Upon admission to the hospital, however, a fracture diagnosis is usually listed rather than osteoporosis.

By analyzing data from a large variety of sources, each with its own strengths and weaknesses, it is hoped that the results presented have been integrated into a coherent and reasonable understanding of the impact of these disorders on the population.

Index

E

F

G

H

N

O

P

R

S